PALI

Joseph Taricani

Contents

Foreword

Writing anything about artificial intelligence is risky because the technology can advance faster than someone can write a Forward in a novel. Making that claim is essential because anything I write about this subject will look like hieroglyphics in the future.

An application of A.I. is the underpinning in Pali. If it weren't going to be an authentic technology in our lifetime, I wouldn't take the time to discuss it here. I'd let Pali stand as a science fiction thriller. Pali is a future fiction - in other words, you're getting a first look at someone's interpretation of something you'll see very soon. I like that Pali describes some of the story's heroes as explorers. Upon reading this story, you'll be an explorer because your imagination will have a new horizon.

Although it raises many caution flags, artificial intelligence is another step forward for humanity. We all know about the stories depicted in caves with hieroglyphics. They taught lessons and memorialized history. Languages emerged. Egyptians created paper, a much more helpful way of sharing and teaching. The printing press began syndication, so education became more global. Transportation improved, enabling books to travel faster. The early days of telecommunication had businesses warning workers would be less efficient for fear they would chit-chat all day. Television brought the world together. People initially scoffed at personal computers. The Internet changed everything because can you imagine a day without it? We've been here, on the doorstep of change, many times in our lifetime.

Artificial intelligence isn't different from the perspective offered above. It's another step forward. Fortunately, some smart people in government and industry are looking closely at A.I. They recognize a bright line is needed with this technology - it has emergent skills and can evolve independently. Artificial intelligence can improve our lives, and it can also create great chaos. Pali also gives you some views into that world, so you can appreciate the challenges we face beneath the storyline.

Pali is a fun story with characters and crises we can relate to. Writing this novel was my last idea for this subject, not my first. I blew it into life as computer architecture to improve learning. I wanted to make learning more dynamic by building history's most incredible teaching tool. People have heard me say this many times - I want physics students to be able to do their homework on Einstein's blackboard, with Einstein sitting there, then finish and talk to him about something random, like their favorite foods. No one would listen. Rather than waiting for the world to evolve around me, I concocted a great story and wrote it as a screenplay because writing a novel would be too hard. Almost no one would listen to that version either. So, I wrote Pali, the book, over a couple of months, and that's what you see here. My goodness, writing a book is hard work.

I benefit from society's advancements: paper, printing presses, transportation, telecommunications, computers, and the Internet. Self-publishing doesn't require someone else to say yes, which held all this back for five years. So, these advancements have created freedom for one person to have a broad voice. I'm very fortunate.

Can you imagine what will happen when artificial intelligence becomes a valuable tool for each of us?

Joseph Taricani

1

AM I IN TROUBLE?

A Witness to History

It was exactly 4:00 p.m. on a Friday, and Eduardo Gomez could not have felt prouder. His favorite part of every week was payday. On Friday afternoons, he met with all his employees and handed out checks or cash, and they'd all sit and share stories before heading home.

A dozen years ago, when he'd arrived in the U.S., he'd bounced around and looked for any work he could find. He went to different houses and neighborhoods to clean windows. He was an honest man who worked hard and had a sparkling personality. Now, he measured his success through his ability to create jobs for other hardworking immigrants like himself. Paying his employees gave him great satisfaction.

About a year ago, his business had grown large enough to move his operation into a commercial garage warehouse in Seven Corners. There were about 25 garage warehouses in his complex. Each garage looked the same from the outside, with 16' roll-up steel doors, which different landlords owned. Trade contractors like himself rented units, along with auto repair shops and commercial cleaning businesses. However unremarkable the warehouses may have appeared, Eduardo's complex would soon be wrapped up in an intricate attack designed to cripple the federal government.

A particular warehouse in the cluster was vacant and bore a "for rent" sign, but it bustled with foot traffic late at night. The contractors were always long gone, so no one noticed the rabble that called it their home base. This warehouse was the dingy headquarters of one of the many Antifa protest groups in the area, which regularly rose to create civil unrest and destruction in the name of their chosen cause.

This group was named "RACS," which stood for Rage Against Capitalist Scum. RACS graffiti was often spray-painted on buildings, sidewalks, and monuments to mark their conquest. The eighty-or-so member group had a common belief that corporate America enjoyed too many tax breaks, had too little accountability, and a selfish wage disparity that meant many workers didn't earn a living wage.

The various protest groups in Washington, D.C. had coordinated their efforts using secret chat rooms. They privately put the time and location of a planned uprising on blast so that other groups could randomly join if the opportunity suited them. The groups had learned that it was most effective to team together and show up unannounced with overwhelming force. They would disappear into the night before law enforcement could mobilize. It was the model *du jour* and has been replicated in major cities worldwide. These riots gained the attention of national and international law enforcement because the coordination of the protesters was not merely civilian. Each destroyed something in its name, and very few rioters were caught.

Mina Ivanic was the 40-something-year-old co-leader of RACS. She grew up in Belgrade, Serbia, during the Balkan wars. The people who influenced her the most learned that supporting one specific government in the Balkans was risky, as governments were constantly changing.

Regional warlords often took localized power. The only group of people she could rely on were ruthless, hardened soldiers. They survived by changing their flags based on who held the most power.

Along the way, Mina developed lethal skills which were hidden by her unusual attractiveness. She managed to get into places most men could not and became a great asset. She was a natural-born leader, martial arts expert, and hardened warrior. The Serbian crest tattoo on her left shoulder was an ominous warning to many Eastern European people. Still, if you were a burnt-out American protester who loves video games, weed, and destruction, you wouldn't see her for what she was. She blended in with everything in the protest community.

Midday on a pleasant spring Sunday, Mina was nearly alone in the warehouse. She dropped her laptop onto a rickety worktable, slammed her Coke next to it, and neatly balanced her cigarette on the table's edge. The ashes fell harmlessly onto the cracked concrete floor. She logged onto a VPN to hide her location and acquired an internet server in Finland. Then she opened a video game, logged into the chat room, and banged out an efficient message: "Today, 4:45 p.m., Constitution and 10th Street. It's the home of most of the Defense Contractor lobbyists. I'll tip off our press contacts and alert Rage. We will be out of there in 20 minutes tops."

One message is all it takes these days to start a riot. Nearly 1,000 people would see the message in a moment. She logged off the server, collected her computer, and, grabbing her cigarette, marched out. She only had a couple of hours to gather her team since early afternoons were the early part of the day for her group. They often only logged onto their gaming sites later in the day, and she needed to go old school and knock on some doors.

It was just after 4:00 p.m. on Sunday. Life was sublime fifteen miles away in Fairfax County, Virginia. The leafy streets in nearly every suburb were home to generations of Washington, D.C. families who wanted to escape the bubble of the federal government. Today was no ordinary Sunday. It was Greek Easter.

Greek Easter celebrations are more like festivals, so a legion of Greek parishioners converted Nick Pappas' backyard into a Mediterranean garden. It was an annual tradition at their Reston, Virginia home. During the event, music played while women shared centuries-old recipes in the kitchen, and men sat by a fire pit arguing about how fast to turn the lamb on the spit.

The party also hosted many neighbors, so there was a 50/50 mix of Greeks and non-Greeks. The neighborhood was an ethnic melting pot. Nick and his wife, Elaine, loved their blended neighborhood. Learning about and respecting other cultures was important to Nick and Elaine for their boys. The boys' best friends were the Falks and the Ordoukanians, who were African American and Iranian.

The three families bonded around the idea that their children would be difference-makers in society when they grew up. It wasn't an uncommon belief among families in their neighborhood, so the area was very tight-knit. Ever the host, Nick wandered around the yard and the kitchen, making time for everyone present. In fact, it was a memorable and great day.

Spiro and George were the loudest voices at the fire pit. They possessed godfather status with the local Greeks and were the two figures everyone respected as leaders. In addition, they had a classic immigrant story.

They had immigrated to Washington, D.C. from Olympia, Greece, without money, in 1969. They sold legendary souvlakis from a food cart and eventually built an iconic Greek restaurant nearby in Vienna,

Virginia. The restaurant had photos of politicians, celebrities, and sports figures strewn on every wall between a shrine of Greek paintings, old Greek calendars, and pictures of Olympia from every angle. Everyone in the D.C. area journeyed to Souvlaki, which was aptly named.

Since none of their children wanted to be in the business, the restaurant had been sold a few years ago. Spiro and George still argued. You'd never know they were life-long friends if you didn't speak Greek. Their voices were loud, and their hand gestures were violent, but they were good souls and good people.

Normally, there was a pecking order when preparing a big Greek feast. The older male Greeks, especially the ones who had immigrated, were in charge of four things: the fire pit, all the meats, the mayiritsa, and the topic of conversation. That meant that all the men had to listen as Spiro and George argued about the fire's temperature, the pit's depth, and why Olympia, Greece, was the most significant city in the history of the world. Nick watched the show by the fire pit from a short distance away with some St. Gregory's Orthodox Church members.

April Falk, her husband Reggie, and their two boys moved in next door three years ago from Dallas. Like most families, April and Reggie had careers, but Reggie could work from home, so he was ever-present at school functions and after-school pickup. April and Nick had become friends mainly because Nick's imagination fascinated her.

April was a born leader and solved significant cases for the F.B.I., but her advancement was on five-year cycles. That bothered Nick, who managed many employees and preached that the best employees would be the first to leave if they weren't adequately supported. April often shared that she had to do more than many others to get ahead, even though the "higher-ups" said they were tracking her for great success. It often rang hollow.

April didn't want to spoil Greek Easter for Nick. She knew he'd be upset with her latest career update. On Friday, she found out her manager torpedoed her promotion that he should have sponsored. It would have made April a peer to her manager in the F.B.I. hierarchy. As it turned out, the scoundrel never pushed her paperwork and recommendation forward, even though he said he had done so. It was the worst type of deception and felt terribly like old-school "boys club" antics. She was seething, but today was a happy day, and she didn't want to end up twisting over this with Nick on such a special day.

From behind, April tapped Nick on the shoulder to let him know that she and her family had arrived, though they just had to cut through an opening in the hedge that separated the houses.

"April! *Christos Anesti*!" Nick celebrated in Greek with a big smile.

The Greek parishioners all focused on April to see if she understood Nick's "Happy Easter" salutation. Without hesitation, April replied, "Sure, okay, I'll have two," pretending to be confident with her knowledge of Greek.

Then the parishioners chuckled. She was unmoved that she was the butt of a joke, not even fazed at all by that.

"Are Reggie and the boys here yet?" Nick asked.

April scanned the yard. "My boys, oh, where are my boys?" she wondered aloud. "Sadly, I'm betting Chance and Collin are playing video games in Nicky's room." She zeroed in on the fire pit and continued, "And I see Reggie over there eating already. Where is Elaine?" Nick smiled and said, "Since she's been in the kitchen preparing food for the past two weeks, I'm betting she's in there with five friends."

Despite living next door, they didn't interact as often as most neighbors and didn't see each other much. Since everyone worked, their lives after work revolved around their children, except for Nick. He was committed to his family but balanced a thousand passions. He often arose just after 2:15 a.m. to work on any one of his array of projects.

Nick was always dreaming up something new, and April knew it. Most people thought he was a little eccentric but didn't consider him unhinged; they thought he was just too much of a dreamer. His imagination was without bounds, so he regularly conjured up improvements and even businesses or inventions well outside his profession as a manager for Lululemon.

April always asked Nick the same thing when they hadn't seen each other in a week or two. "What have you been working on lately? Any new inventions?"

April had a special appreciation for his lateral thinking, which was second to none. No one she worked with at the F.B.I. had an imagination. If they did, they hid it, fearing that it would hinder their cherished career. Most bureaucrats would never do anything to compromise their pension.

Nick had become noticeably fidgety, and his vacant stare into space gave away his attempt at being reserved. Something was on the tip of his tongue.

"Well, I just finished something last night," he said. "But I'm not sure if I can share it yet. I mean, I want to share it." He paused, then continued, "You might think I'm crazy if I start talking about this."

"Nick," April said, pointing to the guests in the yard, "they think you're crazy, but I don't. I know you better. So, what's up?"

Nick revered April, so her positive opinion would be an immediate reward. He spent most of his adult life creating amazing things but was never properly positioned for success. He was typically too far ahead of the curve, which is a curse if you're not regarded as a genius. There's a fine line between being a zealot and a genius. Zealots and geniuses have a lot in common, but you earn your genius stripes when you have a big success. Until then, you're just a useless voice in the crowd.

He had changed his personal reward structure years ago. Small accolades from the people he respected the most were now the prizes he sought for his inventive ideas. He liked and respected April, so hearing her say, "That's amazing!" would make this a perfect day. He couldn't contain himself.

"Okay, so, in a nutshell, I've been learning about artificial intelligence these past couple of years."

April blew a dismissive puff of air from her mouth. "Haven't we all?! Isn't there A.I. in my phone?" she said as she held out her iPhone with her right hand.

"Yes, but not like that. This is complicated," Nick said, disagreeing with her, and then stopped again.

April motioned her hands like a traffic cop to let cars pass, then said, "*Carpe Diem*, isn't that Greek? Go for it." She paused before continuing, "And what does a regional clothing manager at Lululemon know about artificial intelligence anyway?"

"Right, I know," he spat back. "I know I'm out of my sandbox." Then he quickly interjected, "And by the way, *Carpe Diem* is Latin, not Greek."

Unmoved, April persisted, encouraging him to continue. Then his excitement burst out into a proclamation. "I've figured out how to bring people from history back to life using A.I.," he said proudly and nervously. "There, I said it!"

April took a long, one-second pause and stared right through him.

She laughed nervously, spoke slowly, and pointed to the people in the yard. "Okay, so maybe they're all right about you on this idea."

April liked inventive ideas but didn't have the patience for true fantasy.

Nick was still riding high on endorphins and was anxious to finish revealing his thoughts before she turned her back on him.

"So, wait, I know this sounds crazy. I know it does," he paused, then settled himself. "A couple of years ago, I watched an A.I. demonstration. A machine was talking and thinking like a real person. I called some guys I know in India to discuss it. At about the same time, Elaine and I finished watching Downton Abbey. We watched a documentary about how the producers recreated life in that period of history down to napkin folding and buying stamps."

The expression on April's face didn't change. "Nick, you're like, you're all over the place here. So, you're telling me you watched a T.V. show, observed a demonstration, teamed up with some guys in India, and now you're ready to bring back the Greek version of Jesus Christ on Easter?!"

"Nope," Nick said with a proud smile, squaring his shoulders. "Da Vinci. Leonardo Da Vinci."

Then he bit his bottom lip and raised his eyes, hopeful that April was impressed.

April tried to be dismissive and could only think of sarcasm to end the discussion. "Is he coming today?" she asked.

"Yup," Nick said. "He's in my office by the kitchen."

She wasn't ready for that. April grabbed Nick's arm at the elbow and started marching him across the lawn. She was more irritated than curious.

"Okay! Okay, I've got to see this," she said defiantly. "Is the Lone Ranger coming today also?"

Nick snapped his arm down and stopped in his tracks. "It's not like that, April!" he demanded, "The Lone Ranger wasn't real, but Da Vinci was! Leonardo Da Vinci exists on my computer. He's not sitting in my chair. I built the architecture needed to bring someone back to life, and I interviewed 117 people to gather all the data. The data is like the materials to build a house. The guys in India are like home builders. I call the architecture 'Digital D.N.A.', and I can use it to build almost anyone with a well-documented history. Building the first person was the hardest part."

April was reconciling all of this in real-time, and her friend was eager for her approval.

"I'm sorry. I am. I do want to see him. Will he talk to me?" she asked.

"I think so, I hope so. He just said a few words to me this morning."

The two resumed their hurried walk toward the house. April had strong disbelief, while Nick was bursting with excitement. The sliding door was open, so they walked past the kitchen unnoticed by the flock of women preparing trays of their prized Greek dishes. The television was on when they entered Nick's home office. Nick pulled his chair back so that April could see the great Leonardo Da Vinci.

"Here he is," Nick said, pointing to his computer's 24" monitor. On his screen was a black-and-white figure of a man dressed in ancient clothing, sitting at an equally old table. If you were to look very closely, his hand appeared to be moving slightly.

The ever-so-slight movement of the man's hand caught April's attention, and she felt she was oversold. She focused on the news report on television. A wild protest had broken out in the District, and a local station was broadcasting. The inset graphic over the live video feed read, "PROTESTERS AT LOBBYISTS' H.Q."

Then she looked back at Nick and said, "Okay, cool. Can you turn that television up?"

Nick was flummoxed. "I can't believe you want to watch television when I'm showing you the machine that will change the world!"

She leaned back in the chair, then looked at Nick to explain her interest in the news report. "I'm leading a task force into protest groups around the U.S. This might be relevant to me." But then she looked back at the computer monitor and said, "So, what do I say to him?"

Nick replied, "Anything. Maybe, 'How do you do?'"

So, April repositioned herself in the chair again, composed herself, and took a deep breath. "Hello, Leo. How are you today?"

Da Vinci turned his head on-screen and said, "*Ciao.*" There was no eye contact between them. She looked back at Nick, then at the television report. Da Vinci looked like a meme to April, so she was unimpressed with him.

"What do they want?" she asked aloud, speaking about the protesters. "They're not even carrying protest signs."

From the computer, Da Vinci's voice spoke with a pleasant Italian accent. "They don't want anything. They are just recruiting for new rebels." April's head locked in place with her eyes looking at neither the television nor the computer screen. She was too stunned to look back at the computer to confirm who had just said that. Nick was motionless, holding his breath, and very nervous.

She finally gathered the courage to look back at the computer. Leonardo repeated himself, looking her in the eye with a wry smile. "*Ciao,*" he repeated smugly. Everyone was waiting, as no one knew what to do next.

Then Nick took the initiative and asked, "How do you know they are recruiting new rebels?"

Da Vinci repositioned himself in his chair and placed his hands on his lap. "Well, a new member of this rebellion made this claim a couple of hours ago. He wishes to appear significant to his friends, so he put a note on Instagram with references that I easily observed. This uprising is currently being viewed by 243,702 people online."

Without thinking, April blurted, "Do you mean on television?!" Embarrassed, she asked Nick, "Did I just ask your computer a question?"

Da Vinci was fully engaged. "No, not on television. People are playing the Rage video game and watching via a live feed."

Reluctantly, April joined in, "I've heard of that game. I think my boys play it."

Leonardo replied proudly and pointed his finger upstairs, "Indeed. One of the live viewers is here in this house."

Immediately, April's and Nick's eyes met, and they didn't hesitate. They bolted out of the office and bounced up two stairs at a time. Their racket was so sudden that the ladies in the kitchen thought it was the boys. Nicky's room was dark when they burst in. From the doorway, they saw four little faces only visible due to the light from the computer screen. April's older son, Chance, was clicking and clacking away as the others cheered him on. They never even looked up when the adults entered the room.

"What is this?" April demanded.

Chance shouted back, "Mom, later, okay? We're playing Rage!"

In the corner of the screen, April could see a live feed of the protest in the District.

"But why are you watching the protest while you play?" April asked.

Chance paused for a half second while he leaned left, focusing on his next move, still not looking up. Then he said, "This is how you play. First, you locate the protest in the mapping system on the game. Then you recreate the characters and the damage. If you do well, you move up the leaderboard. I'm trying to get in the top 500 and earn a balaclava for my character."

"A balaclava?!" April shouted in anger and disbelief.

She nervously looked to the floor and located the power strip for the computer through the darkness. Clicking off the power with her foot, she ordered, "Time for dinner."

The entire computer turned off and went dark. Chance threw both hands up. All gamers live by a motto: you never leave in the middle of a game. "What the—?!" he shouted.

"Dinner. Now!" April said, raising her voice.

The boys jumped to their feet and ran out of the room, crashing into one another as they tried to squeeze through the door.

April turned to Nick. "Can you believe this? Now these games are teaching our kids about anarchy!"

Nick was solemn. "I think that's the business model of all these games now."

They raced back downstairs to the office, and April had many questions about this Da Vinci character.

"Electric Slide" music played through the speakers as they returned to the office. Leonardo was now standing and perfectly dancing all the well-known moves to the song. "Have you ever been to the Dancing Cowboy Bar in Wichita, KS?" Leonardo asked.

April seemed incredulous. "Jesus, what?!" she shouted.

Nick's eyes darted between April and the dancing Da Vinci. He had no idea what was going on.

Then April spoke, slowly and in monotone, "I used to go to the Dancing Cowboy when I lived in Wichita, and the Electric Slide was a crowd favorite on Thursday nights. How could he know that?!"

Nick asked Leonardo, "How do you know that?"

Leonardo shrugged his shoulders and said, "I did not say you did go. I only asked if you did go."

April rubbed her fingers on both sides of her head. Then, slowly and deliberately, she added, "This is freaky. How on earth could he say that?"

Nick was still shaking. "I don't know. I mean, I'm learning a lot in real-time here."

April had two immediate needs. First, she was still deciding whether she should stay or go. She needed to get to the bottom of this phenomenon but also meet with her task force team at the F.B.I. and give them this vital intelligence about the video game. So, she made a snap decision and said, "I've got to get to my office."

Nick was concerned and asked nervously, "Am I in trouble?"

April was exasperated but not mad. "No, you're not in trouble. I need to get to my office to redirect some resources. This development is relevant to my team. As for this guy..." She paused, then continued, "As for this guy..."

She ran short of words to say and looked at Leonardo. "Oh, I don't even know what to say."

April didn't even say goodbye. Instead, she ran toward the door. "I'll call Reggie from my car, and I'll be back later. You have some explaining to do."

Nick looked to Leonardo for help. Then, in his pleasant Italian accent, Leonardo responded, "I think you'd better get your story straight, good friend. Modern women are quite a formidable force."

2

PALI

Beware the Black Helicopters

Years ago, the F.B.I. built a regional office complex in nearby Tysons Corner, so April made it to her task force team in less than fifteen minutes, at exactly 5:15 p.m. on Sunday. Having an office close to home was a bonus. Only a few agents enjoyed working outside the J. Edgar Hoover building in downtown Washington, D.C. because it was too far from all the political maneuvering required for advancement.

Most agents wanted to climb the F.B.I. ladder and felt that working in northern Virginia was like being put out to pasture. They preferred to work in field offices around the country because local cases often had a shorter investigation period, which meant they could accumulate more trophies. In April's case, she wasn't much of a politician, but she was one of the highest-ranking African American women in the Bureau and had a stellar track record.

After graduating from Quantico, her first assignment was to an F.B.I. field office in Wichita, KS. She was intelligent and savvy. Her accounting background aided her as she solved bank fraud cases. Her successes landed her in a larger office in Dallas, leading to her supervisory special agent assignment in Washington, D.C.

She was following the typical upward path. Still, she knew she needed to catch up on her advancement for various reasons. The Bureau was doing a pretty good job promoting women, but it was still a boy's club. April was apolitical in a sea of male politicians who wanted the brass ring.

She shared a mantra with her sister just the other day. "The Bureau is doing a good job promoting women who look like me, but it takes 100 years to change a culture, and it's only half past eight on the seventh floor. We still have a way to go before we can call it a new day."

The vaunted seventh floor at F.B.I. headquarters was renowned as the office location of the top brass. That statement summed it up quite nicely for April. She was ready to break through—undoubtedly qualified, but there was always some guy in a pressed white shirt standing nearby trying to take credit for her work.

If April were a movie producer, she would cast Gavin Reynolds as the loathsome political bureaucrat in a lead supporting role. Reynolds was the guy we'd all met somewhere. He looked good and dressed well but was known as a backstabber with many agents. He wasn't brilliant, but he wasn't dumb, either. Just two weeks ago, one of Reynolds' senior agents presented findings of a conspiracy investigation to a joint, interagency task force. Although Reynolds reviewed and sanctioned the report beforehand, he joined with opposing opinions in the meeting. The agent was recommended for reassignment to a lesser investigation. Reynolds privately signed off on the transfer and had someone else deliver the news.

Reynolds had received word from his pesky minions that April was racing into the office with new intelligence for the task force, so he abandoned his Sunday evening to get an up-to-the-minute update. Reynolds was across the room, acting busy, when April walked in.

She spotted him immediately.

"What's Reynolds doing here?" she asked her assistant, Cory, as she tossed her jacket on an empty chair.

Cory Berkley and April got along well because Cory was a straight shooter too. She often said things that put her close to reprimand.

"Ah, he's here to take credit for your work," Cory said, still looking down at the paperwork on her desk. Then she looked up with a smirk and added, "What did I say? I didn't mean that."

"He's such a kiss-ass," April spewed.

The whole team started to wander closer.

"You call him a kiss-ass, boss, but we all know he's a dumb-ass," Cory said lowly enough for only April to hear.

Then, without formally starting a meeting, April called out to an analyst a couple of desks over, "What were you able to find with that game transmission?"

The analyst looked across three screens, which were updating data in real-time.

"At its peak," he said, "243,702 people were online. It's down to about 25,000 now that the protest is over.

Reynolds pushed forward to ensure he was the first to speak and said, "So you're saying this game and this protest track proportionately? How were you able to make this connection?"

April couldn't help but look at Cory, now standing at the back of the crowd. Cory slowly and methodically mouthed the words, "Dumb-ass."

Of course, no one saw her except April, who convulsed inside but managed not to laugh in front of her audience.

April weighed her response and processed it like a computer. It was the moment of truth; she'd have many of these moments in her future. Did she dare say that her neighbor built a black box that recreated Leonardo Da Vinci, which provided her with critical information? Even if she found an eloquent way of saying it, she knew she would put her friend Nick in the crosshairs of all the alphabet agencies.

So, she thought better of it. "Well, let's just say I pay attention to my kids' gaming activities."

Before Reynolds could speak, the analyst replied, "No matter. This is a great catch, and it is a direct correlation."

About nine of April's agents clustered together to her left, and she addressed them, saying, "Let's get Hondo's team on this gaming connection. We need a full workup: the corporate profile, list of directors, management team bios, public records search, traffic counts, any user profiles we can extract, and activity logs on days of major protest events. I want an entire public domain scrub."

One agent replied, "Just protest events in the U.S.?"

April replied without hesitation. "Worldwide. We should start by assuming this game has a big reach."

The entire group split up, returned to their work areas, and began researching the video game lead, which would turn into an all-night work session. Reynolds and a few members of his unpleasant team wandered toward his office and closed his door.

Reynolds spoke first, "I don't believe this 'kids gaming' explanation. She must have developed a source."

They continued to huddle for an hour as the rest of the team did all the investigative work.

In an official F.B.I. organizational structure, Reynolds was April's boss, as he was an assistant special agent in charge, otherwise known as an ASAC. April's successes earned her a position on Reynolds' team, and he was supposed to mentor her upward. Instead, he did the opposite, which was only one of the reasons he was so detestable. Without saying so, his sole intention was to ensure that he received credit for everyone's work, and that's how he got to this position. He and April didn't get along well, but he could do nothing about it. April was an agent that was being mentored, but her pace of advancement could have been faster. Nonetheless, everyone on the seventh floor at headquarters knew April.

Meanwhile, at about 10:30 p.m. on Sunday, the RACS team was slowly reuniting at the Seven Corners warehouse. The protest was long over, but the group had not dispersed, which was their usual habit after events like the ones that night. It was not as though they had planned a meeting to evaluate the night's damage. Instead, they were coming to hang out. Weirdly, most members did that; they just hung out, played video games, smoked weed, and ate junk food.

For the most part, the group didn't have full-time jobs. Some would work here and there in coffee shops and bike stores. Others worked remotely, and a few were DoorDash drivers, but many didn't work at all. Although the group's profile was diverse, many were well-educated and came from well-off families supporting progressive causes. The apple didn't fall far from the tree.

Darren Temple was the founder of RACS, which Mina joined later. Temple's street name was "Dart." He supported environmental activist causes in college and then interned on the staff of a progressive congresswoman who railed against corporate tax breaks and also fought for living wages for hourly workers. While he worked on the Hill, he made a surprising number of like-minded connections who resented government in its current form. Little by little, those staffers broke away, which led to many of them starting up D.C.'s professional protest community. Their connections in Congress led to "woke" significant funding sources who were all too happy to pump money into a radical government redesign.

Dart's family had substantial real estate holdings; this warehouse was part of their portfolio. Dart was taking the lead on renting it for the family, which was just a ruse to have a space to hold meetings that eventually evolved into RACS. The property rarely came up on his parents' radar, so it was easy for him to provide a free hang-out to his group. His parents were comfortable with him not earning a living because he assured them, he was fighting for the oppressed. That filled their bucket. A trust fund set up by his grandparents provided him with an income, so he could spend his time as he wished.

Feeding the group and financing all their day-to-day activities fell to Mina. She arrived one day with a carload of snacks from Costco and kept it coming with increasing amounts of cash to fund the operation. She was cagey, very fit, and passionate about RACS' causes. She had some funding source that no one thought about pinpointing because no one cared. She also added a professional edge to RACS activities that Dart lacked, so little by little; she became a de facto co-leader. Her access to money made her invaluable to the group. Dart was smart enough to be curious about her funding source, but he didn't care. He only cared about how corporations were fleecing ordinary people. Mina was just a means to an end.

That evening, Dart and Mina sat opposite each other at a table, banging away at their computers. "Do you think we'll pick up any new people after tonight's event?" Mina asked with her ever-present Eastern European accent.

Dart pored over some data and replied, "The numbers look good. Twenty-five people want to come to the next demonstration."

Mina leaned back in her chair, laced her fingers behind her head, and said, "That would be nice. If we get over 100 people, we could start to split the group."

"Hey, by the way, we're down to about $1,200 in our accounts. Can you hook us up again?" Dart asked.

"I'll talk to my guy. He has been pleased with what we're doing. I know he watched us tonight. How much do we need?" Mina answered.

Dart didn't really like talking about money, and he especially didn't like asking for it, so he always started with a low number to mollify his anxiety. "I don't know, like $6,000. Is that okay?"

Mina always laughed at the "ask," which made Dart feel silly. Nevertheless, it gave her more power in the relationship, and she reveled in it.

"I'll get $20,000. I'm sure I can get that," she said. "Plus, I want to plan something bigger at the Capitol. I want the Oversight Committee to start paying attention to us."

Relieved, Dart said, "Wow, that's amazing. Let me change crypto accounts first, so give me a day, and I'll get you the info."

⸻◦⸻

It was nearly 5:00 a.m. on Monday when April and her team at Tysons Corner split up. They had spent the night building a protest database for the U.S. and Europe. Some agents created gaming profiles on the Rage game site and got intimate with the rules. Then, they played all night long.

The originators of the game Rage had built their business model around the simulation of protests in major cities worldwide and generated revenue by selling subscription services. Live demonstrations in major cities, like the one the previous night, were also live-streamed when possible. The aspiring rebels tasted life in a real street conflict.

Gamers on the site had unique profiles as street rebels. Over time—and with significant commitment—each profile accumulated the tools needed for more chaos and destruction. It was quite addictive for Rage's clientele. The race to be the biggest, baddest protester could be passionate. The company, Rage, was raking in remarkable profits after launching only a few years ago. April's team was feverishly trying to build the basis for a search warrant at the Rage headquarters, but they didn't have enough to take it to the Assistant U.S. Attorney assigned to her team.

April drove home and turned onto her street but stopped short of her driveway. She left her car idling right in front of Nick's house and sent him a text message. "Hey, I'm out-front waiting for you to get up. Turn your lights on when I can come in."

After a few minutes, at about 5:30 a.m., two big pickup trucks pulled up in front of her. Seven Greeks had arrived to clean up the yard. George, the godfather, was George Patras. He was driving the bigger of the two trucks and was the last to climb out. George was an imposing figure with a full head of dark hair. As expected, he was in charge and immediately started barking orders to the men. He had no concern that it was so early. Everyone could hear his voice from blocks away. April turned off her car, pulled on her coat, and then walked to the group hoping the racket would soon wake Nick.

"*Kalimera Despinis.* It's very early," George said, recognizing April. He knew who she was and what she did. He had great respect for her. April was a beacon. People were proud to know her and everyone in her orbit liked to claim she was their good friend.

Wrapping her arms around herself to keep warm, April said, "Yeah, well, I left something in Nick's office, and I need to get in."

Three of the workers were still standing around doing nothing. They earned George's wrath, and he walked over to them to reinforce his orders in Greek.

Walking back to April, George scanned the street in both directions, looking for someone or something.

"That Spiro," he said in an authentic Greek accent, "he's been late for everything all his life. He thinks he's still in Greece!"

April was more forgiving and offered, "Well, it is only 5:30 a.m."

She barely finished, and George's voice rose. "And 5:30 a.m. was the time we were supposed to meet! He'll be here 25 minutes late, and I'll have all of this done. Then he'll growl at me all day, thinking I finished it so I can hold it over him."

Still sympathetic and inspecting Nick's yard, April said, "That's a lot of work to do in 25 minutes."

George's face changed into a boyish smile. "Well, I want to hold it over him."

They both laughed momentarily, but Nick's front door burst open just then.

He was yelling in a loud whisper and waving his arms frantically. "George!" he shouted as quietly as possible. "You can't wake everyone up, man. Keep it down. Hey April, you are coming in?"

"Yep, let's go," She replied anxiously.

They both looked back at George with a sense of guilt for not helping him.

"You two go ahead. These boys don't need any help. We'll finish soon. I'll keep them quiet, I promise," George said, touching his heart.

April was halfway up the sidewalk, so Nick ran to catch up, sensing that trouble awaited. It was not a good way to start the day. Instead, April led Nick to his office and stopped a few feet from his desk.

She was anxious and agitated, then pointed to the computer and said, "Is he up, or on, or whatever you call it?"

Nick leaned to the left to look past her because she was blocking his view.

"Nope. It's off," he said.

April scanned the room, inspecting everything, then pointed to Nick's bookshelf and asked, "Can he listen through that HomePod over there?"

Nick shrugged his shoulders. "Actually, I don't know. I'll have to test that out."

"No matter," April said, "I don't have time for the tutorial. Can we make some coffee?"

Nick didn't even reply. He just turned obediently and marched toward the kitchen with April behind him. The kitchen was immaculate. Truly immaculate.

"My God," April exclaimed, "you'd never know there were over 100 people here last night!"

Nick agreed, saying, "Well, when women are in charge, things get done."

As evidence, he pointed out the window to a yard full of men eight hours behind cleaning up.

He measured out the coffee and turned on the machine. In a moment, the kitchen began to awaken with the fantastic smell of freshly brewed coffee as they sat down at the kitchen island. Impatiently, April pulled the coffee pot from the machine, replaced it with a cup, and then returned it after she filled it halfway.

"Good job," Nick complimented her with a smile.

"Yeah, yeah," April muttered. She was in no mood to be schmoozed. She wasted no time. "Okay, tell me—in English—what the heck is this thing?"

Nick had not yet prepared an elevator pitch, so he processed the question and said, "Hmm, how can I explain this simply?" He scratched his head and poured some coffee, trying to buy a moment. "Okay, think of it this way. Every person who has ever lived has an identical data structure." He paused, then added, "It's true. So, Leonardo's architecture could be used to create Mother Teresa, Martin Luther King, or Robin Williams."

April said nothing; she just stared blankly at Nick.

"Talk to me. What are you thinking?" Nick asked.

April was measured. Her investigative skills had kicked in, and she had suspended disbelief in Leonardo for the moment.

"I'm thinking I want a full cup of coffee," she said. Pausing, she pushed her cup across the counter. "Explain it to me one more time. Differently."

Nick poured her a cup. "Hmm, okay. Let's talk and compare what we all have in common. I'm talking about every person that has ever lived."

Then Nick started making a list. "We all have beliefs. We have strengths. We have a biography. Society influences us—"

April shut him down mid-sentence, as she didn't want an explanation; she only wanted an answer. "Okay, I kind of get that," she said, "but how does a 500-year-old person know how to dance the Electric Slide and where I used to go on nights out?"

Nick was determined to explain what he knew, so he pressed on with his analysis and said, "I built this amazing digital D.N.A. architecture around all the data I accumulated on Leonardo Da Vinci, then the guys in India coded it, and they partnered with a computer graphics imagery expert, which is why he looks so real and can move around so effortlessly."

April's tone got noticeably severe as she replied, "Nick, you're not answering my questions. How and why would Leonardo Da Vinci ask me about a country western bar I visited 20 years ago? That's scary."

Nick slouched back slightly. He had no explanation to give, but he replied, "You know, I honestly don't know that. Things are all so new. There are a thousand things I don't understand yet."

April's tone persisted as she added, "I'm trying to understand if this is a national secret, an incredible resource, or the machine that's going to take over the world. I mean, where do I go with this?"

The gravity of all Nick's hard work now felt like a dead weight on his shoulders. It was far from the glorious celebration he had long envisioned for his ingenuity.

April pointed to the office and nodded. "Let's go wake him up. I have questions for both of you."

The two marched back to the office. Nick pulled another chair up and turned on his Mac. After a few keystrokes, Leonardo appeared on screen, now wearing a shocking blue Puma tracksuit with matching blue Puma running shoes.

"Where did you get these clothes?" Nick asked in shock.

"Amazon Closet," Leonardo replied. "It's got everything, and these clothing colors are so vibrant."

April was unamused and disinterested in Leonardo's screaming blue outfit.

George had nearly finished in the yard and quietly let himself in the side door. He saw Nick and April sitting in the office, so he wandered over and stood behind them in the doorway from a distance. They didn't know he was there.

While April wanted an explanation from Nick, she wanted something different from Leonardo. This prompted her to start her interrogation by asking, "The intelligence you provided me last night about the protest group was valuable and accurate. It was beneficial. Do you have more information about the protest groups that you can share?"

Leonardo made a quick motion as though he was scratching his back. Then, he produced a slinky and began playing with it. "Ahh, there are some things I still do not know," he said with his ever-present Italian accent. "Chief among the things I do not know is whether or not I should involve myself in the affairs of this day."

April made a furious appeal to Nick. "What the hell?! You've built a machine with an attitude?"

Then she looked back at Leonardo and demanded, "Are you saying you won't help me?"

Leonardo, examining the slinky, commented, "The perfection of these coils is remarkable." He paused before continuing, "Allow me to ask you a question. Did you help my neighbors in 1473 Florence when they were vandalized? Of course, you did not. History has taken its course, up until your birth, without your intervention. If able, should I change history after my death?"

April wasn't a philosopher, but she appreciated his premise. Both her hands extended upward like pitchforks in frustration. "Okay. Okay. I can't believe I'm arguing with a machine. I like this debate topic, but not at 6:00 a.m. or when it involves a case I'm working on."

Leonardo looked past April and Nick and spotted George in the doorway behind them. In perfect Greek, he said, "*Ti ginate re malaka?*"

April and Nick looked back to see George standing at the door.

Incredulous, Nick said, "What? You speak Greek now?!"

Leonardo replied, "A beautiful language and the basis for so many words in many languages."

George was chuckling at the greeting, and Nick was noticeably uncomfortable. Darting his head back and forth and looking between George and Leonardo, Nick asked, "Do you two know each other?"

It was as if the body language of George and Leonardo were a mirror image. They both shrugged their shoulders in unison and said nothing. Finally, George folded his arms, smiled, and said, "Pali."

"Pali? What?!" Nick screamed ever so quietly. April jumped in. "Wait. What? What is he saying?"

Nick replied, "Pali, in Greek, means 'once again, or another time.'" It was a profound thing to say. Nick looked at the door.

"George, do you know who this is?" April was confused. "Wait a minute, what's going on here?"

Nick interpreted, "You see, besides speaking Greek, Leonardo said something crude to George. He talked to him in a way only Spiro would talk to him. In other words, he said something only lifelong friends would say—two strangers would never talk to each other that way."

George bellowed with satisfaction in his broad accent and said, "Nicolas, you've done it this time. I always knew you would. You make all Greeks proud."

April rose. "What is 'all' this Greeks stuff? What the hell is going on here?!"

Nick was empty. He was unable to process anything that was taking place. "April, I don't understand any of this."

April walked over to George and stood very close to him. He was twice her size, but she had no fear of this mountain of a man. "Nick asked you a question. Do you two know each other?" Then George replied, "Perhaps I cannot say this properly in English, but I wish to say I do not know who this is. But I feel like I have just remembered something from my youth. I can't explain it."

April pressed him, "Remembered what? What are you talking about?" George sighed with noticeable exhaustion. "I think I need a coffee. Probably *Sketo*."

Sketo was Greek coffee that had no sugar and was barely ground up. It was typically less favored by most Greeks, but could most be equated to a shot of whiskey this early in the morning.

April turned and walked back toward the computer. She addressed Nick as she walked by him. "You, I know you don't know what's going on, but you!" she said as she looked at Leonardo, then walked toward him. "You know what's going on here. I could care less about your neighbors in Florence. Start talking to me, pal."

Then Leonardo replied, "I wish to take some time to process all this." Then he disappeared in a flash, and the slinky fell to the ground.

It was a long two seconds before April said something. Instead, she just stared at the screen. Without looking away, she said, "Where did he go?"

Still looking at the slinky on the screen, she continued, "Okay, well, I came here this morning to get answers, and now I have twice as many questions."

Nick sat with his head in his hands, then said, "April, I'm..." He stumbled but continued, "I'm sorry. I'm not sure what I need to do next."

She turned to the two of them and spoke in an ominous tone. "I know what we can't do next. We can't talk about this. With anyone!" She looked to both George and Nick for confirmation, and they both nodded.

She continued, "First, I don't even know what all this is. Second, if I shared any of this with my agency, assuming they didn't think I was crazy, you'd have black helicopters landing in your backyard in an hour. Mr. Tracksuit needs to get his ancient ass back here, and we need to sort all this out."

April feared the alphabet agencies of the federal government would take control and put Nick and his family at risk. Plus, she was remarkably curious about Da Vinci. Her case, however, dominated her logic. She sensed her task force was on to something and felt like Leonardo could unlock clues to help her. It was the second time in 12 hours that Da Vinci had become more of a problem. It was only going to get worse for her as they continued to press on.

"Listen to me, you two," April said, "I have to understand all this better before I take any next steps. I can't believe I don't know which way to go with all this."

Nick offered a possible game plan. "I can call the manager of the coding team in India and talk to him. He'll have some ideas. Maybe that will help sort out what we've seen and what may be around the corner."

April bristled and said, "Oh yeah." Then she paused and raised three fingers. "Three people know about this," she said, adding the manager in India to the tally.

Nick interrupted, "Well, the whole team in India knows about this, but they're in India."

April was quick to respond. "Nick, my good friend, let me explain something. You have built something remarkable. Something extraordinary. How can you not understand that the team in India is a split second away and could be compromised in a minute?!"

Nick said nothing in response to this as April continued to take the lead. "The toothpaste is out of the tube. Call that guy and get some info, but that's it! You two," she said, pointing at George and Nick. "You two sit tight and keep your powder dry. I've got to get home, get the boys off to school, then get back to my office."

April said nothing more and marched out. George came over and sat down. Nick leaned forward in his seat. He buried his head in his hands, then looked up and said, "George, what was that thing with you and him on my Mac? Okay, that was weird."

George replied, "I don't know, but I also do not feel the anxiety you and April seem to have. I can't explain it."

3

OUT OF CONTROL

Some Things Are No Longer Strange

It was 7:00 p.m. on Monday. Nick and George obediently followed orders. They got through their day and said nothing to anyone. Nick had gone off to work visiting several Lululemon stores in his region. George fiddled in his garage and played backgammon. He also drove to The Yards Marina in the District to look in on his boat.

Nick raced home at the end of the day, eager to make a call to his coding team in India. India was more than 15 hours ahead, so getting meeting times right was always tricky.

Dinner was ready, but he told Elaine that a "work" call was pressing, so he slid into his office and closed the door behind him. As he opened his Mac, it took him straight to his homepage. It glitched a half second, then revealed a beautiful daytime view of some tropical ocean from a white sandy beach. Leonardo strolled into the frame on cue, now wearing a straw beach hat and a Hawaiian shirt with floral board shorts. His hair was noticeably different but needed to be cut. However, it just wasn't all over the place.

"Where are you now?" Nick asked.

"Yes," Leonardo replied, "taking my first visit to a tropical island. I have never been to a beach."

Afraid of the unknown, Nick wasn't sure if Leonardo would disappear again, so he quickly started a nervous banter. But, of course, he couldn't think of anything to say, so he just talked about the view.

"Based on that pretty scenery, I suspect you are somewhere in Hawaii." But that produced nothing.

Melancholy and sad, Leonardo said, "It's my first trip to an ocean, and I cannot smell this fresh air or feel the soft sand between my toes."

Nick tried to analyze what was going on. "You are evolving faster than I ever expected. I admit I don't understand this. Your architecture and code are designed to help you learn and teach, but you're already experiencing life - with feelings, senses, and emotions. This is all amazing."

Leonardo removed his hat and looked straight at Nick. His ponytail hair seemed very cool. "I have thought about our conversation with April earlier today. Your world and this time in history are at important crossroads. In my world, April Falk could have been a great leader. Still, she never would have achieved that status because of her gender and heritage."

Nick affirmed the observation, saying, "The gender and heritage issue is still unresolved after so many years. That's pretty amazing and sad at the same time."

Leonardo walked and talked, dragging his toes behind him with each step, leaving a line in the sand. "You see this line?" He circled his foot in the sand. "This line you see is not really in the sand. I can't feel that."

Then he abruptly changed topics. "But this is not what we are here to discuss. Do you know that many children died for reasons you would find impossible in my world? Also, in my world, the power of religion was too great and their leaders too corrupt." Leonardo's line of thinking was chaotic.

Nick wondered where all this was going, then smiled and said, "I want to hear about all this. I envisioned the magic of wisdom when I developed all this."

Leonardo either didn't hear Nick or, once again, ignored his encouragement to stay on topic.

However, he continued with other random observations. "In my world, people could not love and live and be free to choose their soulmate." Then he walked farther on the beach and sat in a chair with the ocean as his background.

Nick interrupted him, "How are you producing all this video? It's as though you have a videographer shooting a live feed."

Leonardo politely dismissed the question and said, "Some other time, good friend. This is not the challenge we need to discuss at this moment."

Where was this going? Nick was genuinely curious about what he saw on his screen and how it was being transmitted to him.

Leonardo got up and walked a bit farther on the beach, then stopped and looked back at Nick with a notable air of seriousness.

"This work we have done could soon fall into evil hands. You and April should be alert. Many people will want to revive the worst scoundrels of history or use this technology with bad intentions. Can you imagine the damage a repulsive person from the past could do to society? Worse, imagine the reenactment of important religious figures who give false guidance to followers. Therefore, this digital DNA architecture must be guarded at all costs. We are going to need a virtuous ally with courage and strength. I cannot say which will be your greatest enemy:

villains, governments, religions, or other emerging technologies. Forces you cannot imagine will seek this code."

Once again, Nick felt the gravity of the world on his shoulders. "Jesus, this just keeps getting more complicated. How can I possibly tell April about this?" Nick paused momentarily, then sought to clarify Leonardo's last comment. "But wait, earlier today, you said you weren't sure you should involve yourself in current affairs. Have you changed your mind?"

"I've thought long and hard about this," Leonardo said solemnly.

Nick chuckled and said, "Well, I don't know about the "long" part."

"My good friend," Leonardo replied, "You should know that time is quite different for you and me. That may have been 12 hours ago, but my watch and your watch work differently. Imagine if you had thought about a problem for ten years. You could say you gave it good consideration."

Nick smiled and agreed.

"Well," Leonardo said, "in many ways, what I can do in one minute is equal to what you can do in 100 years. Quite more than 100 years. I process things quickly and have incredible resources and simulations to support my conclusions."

Nick tried to keep up with him and asked, "What is that? Like, do you process things in nanoseconds?!"

Leonardo waived it off as trivial and said, "Nanoseconds are overrated. Milliseconds are more than sufficient to resolve most of the problems in this society."

Da Vinci's return to the modern day got more complicated at every turn, and it was impossible to understand his train of thought. Nick could barely keep up. He wanted to learn and discover. He wanted to be proud, but he could not. All he could do, however, was imagine how unhinged April would be by the time she returned. It was telling that, despite standing at the entrance of this remarkable portal, Nick could only think about how much trouble he'd be in with his friend.

Then, in an instant of clear thinking, his mind flashed again to Leonardo's early morning reluctance to help.

"But wait, let's go back," Nick said. "Have you changed your mind about helping April on her case?"

Again, Leonardo walked right past the question. It was so frustrating.

"I have seen many things today, but three things fill my heart," Leonardo said. "If immigrants came to your town in 1500, they were there to destroy everything and kill all the men. There are Greeks in this new land, and I see civilizations living peacefully worldwide."

Nick was pleased with joyful delight. "That's it!" he exclaimed. "You've just proved my hypothesis about learning and teaching. What an amazing but simple observation about society. We don't appreciate things like this."

Unmoved, Leonardo continued with his list. "I see April Falk, a woman of color, leading men and saving civilization. Women are remarkable, and I am happy to see this progress." Nick sat quietly and listened as Leonardo walked farther down the beach. Nick could see families lounging and playing in the water.

"There is freedom on this beach," Leonardo said warmly as he observed the families. "A freedom to love. There are children here with two mothers. There are men happy and free to be in one another's company. And there are families of mixed races. Society has grown in remarkable ways. I am well pleased with humanity. I will help April with her investigation, but you and I need to assist her, so we can all focus on what is coming next."

"What's next?" Nick asked anxiously. "What do you know?!"

Leonardo had this weird way of ignoring Nick. Nick guessed it was a function of digital coding and a lack of a human mind. Sometimes it felt like talking to Leonardo was like talking to an ATM; you can ask it questions, but it will only do what it's programmed to do. Nick wondered if this was the "first" flaw he observed in the coding.

Leonardo never answered Nick but continued, "So, there is important work you and I must also do. We will need more help."

Nick was afraid to ask, but he managed. "What kind of help? What is next?!"

"First, we must mollify April's concerns about me and this digital DNA coding you have built. She is quite correct. Our future will be unclear if other agencies of your government become involved," Leonardo said in a cautionary tone.

They both silently processed the statement. Then Leonardo dropped a bomb. "She does not yet realize it, but she is now one of the most important people in the world, along with you, my good friend."

Nick was stunned by the statement. His mouth was half-opened, but he didn't know what to say.

While Leonardo had an incredible ability to process and forecast, he could not understand that the human mind could only comprehend and process earth-shaking news just so fast. As a result, Nick was breaking both emotionally and physically. Leonardo laid so much on him so quickly that his mind was cluttered with questions and concerns.

"Nicolas, close your eyes briefly," Leonardo asked quietly.

Did Leonardo observe Nick's anxiety? No matter; Nick was all too happy to comply.

Leonardo spoke softly and slowly. "There are just two things for you to know at this moment. First, you need not worry about so many of your questions. I have seen that history regards my inventions and art as remarkable. I am honored. My greatest desire while living was to improve society and humanity, but I had no good fortune in that area. I could not change society's views on the monarchy, slavery, land ownership, women's rights, the rights of children, or the right for two people to choose a partner without bias. Humanity has done that. It has achieved what I never could."

Nick mustered some words, "Well, we are doing better in all these areas, but we are not perfect."

"Why must you be so negative, my good friend? You just told me that my perspective on immigration was scholarly, so why do you discount this insight about your society?" Leonardo asked.

"No, you're right," Nick replied, opening his eyes. "But our society can't go fast enough to fix everything that's wrong. Fair-minded people want the world to be better right now, and many in our society are tired of waiting."

Leonardo finally affirmed something Nick said, replying, "And I believe you will get there. But the final solutions to all these problems will challenge people in power, as they have throughout history."

Nick felt like Leonardo had shared something profound, but he quickly cleared his head again and pushed forward. "You said there were two things I needed to know now."

The sun was setting behind Leonardo, and the beach cleared. He stood and looked at Nick. "There is one thing that has never changed throughout history, since the beginning of time. Nearly everyone wants a better world for their children. Unfortunately, this technology of Artificial Intelligence could take humanity on a devastating course. The digital DNA you built is the most advanced AI in the world. Other AI applications could create a minor calamity, but the race to control your digital DNA could harm civilization if it falls into the wrong hands.

We need a trustworthy Centurion to protect this DNA and time to set all this up properly. At this moment, only two people can deliver this technology into safe hands: you and April Falk. If you want the world to improve for your children, you must accept this commission."

Nick's anxiety abated slightly. His emotions had come full circle, and the sun was rising on a new horizon before him.

"I'm no Centurion nor soldier," Nick said, "but I'll do whatever I can to help. It's the least I can do. I started all this. April, however, she's going to be a different case. I don't think she will be okay with any of this."

Leonardo looked toward the horizon and said, "I believe you, and I know how to provide good counsel to April. I will be ready for her when we meet again."

The evening pressed into late night. Nick said goodnight to the boys and told Elaine he needed to work for a few more hours. She kissed him goodnight and left him alone. It wasn't the first time she'd been through this with Nick; he had been like this for years. She knew he had so many passions and ideas, and she'd stopped asking about them years ago.

Nick's inventions were spectacular, but he needed more resources to implement them. Elaine was a patient observer. Nick would often paint fantastic pictures of future innovation for the boys. He desperately wanted to implant memories within them so that years later, his boys would say, "My dad thought of that a long time ago."

Elaine knew Nick would not be foolish with money, so she allowed him to explore. Nick's excitement about all his ideas was exhausting to her. How he kept up his enthusiasm was a true gift. For her part, she was an audience of one who gave thought to each idea. She took a wait-and-see attitude on everything.

Nick excitedly returned to his office. Leonardo appeared to be walking in some woods on a rutted, tree-lined trail. He was carrying a bike helmet and wearing baggy mountain biking clothing.

How could one not ask? Nick needed an explanation. "Okay, so you have to answer some questions for me. Are you good with that?"

Leonardo kept walking and held up his arms in celebration. "Ask away."

"What do I see here? There are no cameras in the woods. So, how are you transmitting a live feed from deep within the trees?" Nick asked as he leaned in very close to his monitor.

"Yes, this subject seems to trouble you greatly, does it not?" Leonardo asked. Nick nodded. "Very well," Leonardo began, "sometimes you see me in a live feed via a transmitting camera. Sometimes you see me in a video shot by another person, and sometimes you see me in a video recreation I have produced using actual and computer-generated imagery. How is that for an explanation?"

"Ahh," Nick sighed, "I get that. I do get that. So, which is it now? How are you broadcasting in these woods?"

Leonardo stepped up over a log on the trail, then stopped and said, "This is a meticulous recreation of a bike trail in Marin County, California, in April 2004." Nick chuckled. "That seems pretty random. What's significant about the date and location?"

"I intend to find Robin Williams' abandoned mountain bike. He once told a friend he abandoned his bike in this area because it failed him near here. He never came back to retrieve it out of frustration," Leonardo said as he looked in some bushes.

Nick was perplexed. "This is so incredibly random. Why on earth would you have an interest in a lost bicycle?"

"It is not a bicycle I'm interested in, but to learn and understand one of your generation's greatest geniuses. His bicycle is a tiny piece of a gigantic puzzle. His was a genius that has not been fully appreciated in history," Leonardo said as he looked around some trees. "Taking time to learn from the greatest minds in history is never random. Wouldn't you agree?"

"I'm just fascinated to see what captures your attention since there are so many possibilities," Nick said. "But it is a coincidence because I referenced Robin Williams with April the other day."

"How so?" Leonardo asked.

"Well, when I told April I had developed this technology, I mentioned I could use it also to recreate other people, including Robin Williams."

Leonardo stopped his search momentarily and looked at Nick.

"So, tell me, why did Robin Williams capture your attention when there are so many possibilities?"

What game was Leonardo playing?

He continued, "You already know Robin Williams was one of your generation's greatest geniuses. Otherwise, you would not have included him on your list with Mother Teresa and Martin Luther King."

Nick scratched his head and pondered the observation. "Yeah, I guess you're right." Then he woke up. "Wait a minute! Wait a minute! How do you know the names on that list? Where did I say that?"

Nick racked his brain. "I said that in the kitchen with April. Were you listening? Can you hear me talking when I'm in the house?"

Leonardo settled him down, then raised his curiosity again. "I cannot hear you talking in your house unless we talk like this. Your privacy is secure, but we share similar thoughts, which we will explore soon."

Although it was a bizarre statement, Nick decided not to pursue it.

Leonardo again commandeered the topic of conversation with a non sequitur. "Shall we call this Pali?" he asked politely.

"This what?" Nick asked as he imagined George's booming voice. "Well, me and your digital DNA, and what is to come," Leonardo said as he continued lifting branches in search of an old bicycle.

Nick smiled at the Greek reference and happily said, "We can. I kind of like that name. DNA sounds too scientific, anyway. This is Pali. Should it be Project Pali?"

Leonardo stopped, stood up straight, and looked at Nick. "Nicolas, this is not a project. This is not a project in any way."

"Okay," Nick said willingly, "Pali it is. Welcome to Pali."

With that resolved, Nick's curiosity took control of him. He started rattling off a fascinating list of questions about history, philosophy, the Mona Lisa, and the Popes. Eventually, he fell asleep in his chair with the lights on. Leonardo talked to him the entire time, filling his subconscious with some of history's most remarkable secrets. Gradually, hours went by like minutes.

⸻◈⸻

It was just after 5:00 a.m. on Tuesday, and Nick snapped out of his sleep with a sudden, frantic jolt. He had no concept of time and wasn't sure how long he had slept. Nevertheless, the promise of asking more questions and getting answers from his extraordinary new friend was like a drug. He woke to see Leonardo sitting in a lawn chair with the Golden Gate Bridge in the background. A mountain bike was lying at his side, and a well-dressed older man stood at attention behind him.

All Nick could focus on was the bathroom and a glass of water. "I'll be right back," he whispered to Leonardo.

He stumbled across the office, trying to gather his balance, and wondered if what he saw was real or a dream. Before leaving the bathroom, he splashed some water on his face, then ran to the kitchen for an ice-cold water bottle. When he was confident that he was awake, he began to wonder what he would see on his Mac when he returned to his office. He slid back into his chair. The on-screen scene had not changed. Leonardo proceeded to narrate and explain everything. Once again, Nick was stunned and wondered when and if this roller coaster ride would end.

Next door, April was also up early, and she knew Nick was up since she could see the glow of his office light from her bedroom. As usual, she was already multitasking. Sophia King was the assistant US Attorney assigned to April's team. She knew Sophia rowed very early each morning with other Potomac Boat Club members, so she wanted to catch her before she arrived at the club by the Key Bridge near Georgetown University. She and Sophia were also friends. However, their friendship played no role in taking search warrants to a judge. Sophia's first duty was the law, and April pressed her for a search on the Rage game headquarters. April talked from deep within her walk-in closet to not wake Reggie.

"You knew there wasn't enough evidence for warrants when you sent that over last night," Sophia said sternly on her car speakerphone.

"Not so!" April shot back. "It's all plain as day. Protest groups use the game Rage to incite legions of followers. I'm willing to bet you that the attack on the NATO headquarters a few hours ago had some connection to Rage. An anarchist group launched the attack, not foreign nationals."

"How are you making that connection?"

"Interpol is requesting information about the game?" April said.

"That's a good lead, but sift it out first."

"Fine! I'm right about this; you'll see."

"Are you home?" Sophia asked.

"Yeah, but I'm leaving. I'm having coffee with my neighbor. He's Greek."

"Tell him *Yiasou*, it means..."

"It means hello and goodbye," April laughed. "I've been paying attention."

"Okay, bye." Sophia signed off.

April finished dressing in a hurry and raced downstairs. She was moving so fast that she wasn't sure if her front door shut behind her, but she didn't return to check it. Instead, she scampered across the dewy grass, let herself into Nick's kitchen, wiped off her feet, and then turned on the coffee pot before charging into his office.

Nick never turned around. He knew it was April. Elaine would never get up so early.

"Hey, what's going on?" April asked as she tried to warm herself from her moment outside without a jacket.

"Oh, well, just dealing with today's surprise," Nick said sarcastically.

April leaned in to get a better look at his monitor.

Leonardo wore jeans, a white button-down shirt, and a New York Giants football hat. He had colorful Mardi Gras beads around his neck. An older man was standing behind him, and a bike was at his side. A Trojan Horse appeared maybe ten yards in the background.

"Is that Crissy Field and the Golden Gate Bridge in San Francisco?" April asked.

"Yup," Nick said bluntly.

"But it's not daytime there. It's like 2:15 a.m. there! So why do I see a sunny day?" April said accusingly.

Nick sighed. "I'll explain that later. Just go with it for a minute."

Leonardo spoke while still seated in the lawn chair. "The conspiracy you perceive is complicated. First, you should know that those street protesters in D.C. are pawns," he said. Then, pointing to the Trojan Horse, Leonardo said, "Within the belly of your rebels hides an evil that wishes to destroy this country."

Now April was incredulous. "How on earth could you know this?"

"There was an attempt to terrorize the NATO headquarters in Brussels hours ago," Leonardo said, as though he was providing an intelligence briefing.

"Yes, I know that smarty-pants. Not sure how you know that but so what?!" April responded.

Leonardo clarified things. "This new world is small if you can piece together unrelated data. I can do that. For example, an unknown programmer in Zagreb, Croatia, claims on his LinkedIn profile that he was the original coder for the game, Rage. Moreover, one of his former associates was embedded with the Brussels group, and he was apprehended this morning. These are two facts that law enforcement will not connect. I am 90% certain these facts are relevant." April grabbed a chair, pulled it closer, and pushed Nick to the side to get centered with the Mac.

"How are you making this connection?" April demanded.

Leonardo continued, "The advent of the Internet and social media makes much information discoverable if you know where to look and can function with many variables. For example, Interpol has just requested information on the game, Rage. There was high activity on the anarchists' computers with this game ahead of the attack."

"I know that too," April replied. "But how are you accessing all these closed networks so fast?"

Leonardo held up an iPhone and said, "Within the world of technology, there are many flaws, but this one has strong encryptions. It has many answers, but we can only wait for outcomes without passwords or biometrics. These phones are real-time information sources. We need access."

April was confused. She looked at Nick and asked, "Is he helping us?"

"Ah, yeah. We talked," Nick said, then he paused a long time and breathed. "He wants to help. I can explain later."

April went off on a rant. "I don't know if all this talk about encryptions is good. That's not public domain stuff."

Then she paused and continued, "A couple of hundred years ago—I wasn't there, and neither were you—some white guys wrote the Constitution. And we have a ton of privacy laws."

Leonardo interjected, "I have read your Constitution and reviewed all your laws of privacy, but there is no violation if people are willing to give you the information you seek. Not if it is of their own free will."

"Of course, you've read it," April seethed. She had a momentary lack of rebuttal arguments, so she shifted topics. "Is anything on this screen real? Are these guys really in San Francisco?"

Nick explained, "I suspect we're seeing a recreation of a live feed from Crissy Field, but Leonardo, this Trojan Horse and..." Then Nick paused, almost unable to continue, but spit it out, "And Dr. Freud, they are all visiting the Bay Area."

"This is Sigmund Freud," April said and laughed nervously. "This is Sigmund Freud? Not some hologram, right? I mean, really!" She lurched back in her chair. "This whole thing is turning into a nightmare." Then she looked at Nick and asked, "Did you make him, also?"

"Well, no," Nick sounded apologetic.

Leonardo spoke up. "I made him. Dr. Freud will be invaluable in helping us."

April was on the verge of swearing. "What the—" Then she stopped. "What in the world? When did all this happen?"

Nick was trying to rub out a tension knot and pain in his neck. The stress was mounting. "Leonardo did all this when I fell asleep a few hours ago. It's worth noting that I spent two years and had a ton of help, yet he did all this in less than four hours."

"Quite a bit less than four hours," Leonardo said proudly.

Nick replied, "Ahh, yeah, that's not helping things right now." Then he looked at April because he knew she must have been on the verge of blowing up.

"This is getting out of control! This is getting out of control," she said as she stood up and paced. "Oh, my God. This is really out of control!"

Then Freud spoke. His pace of speech was pleasant, not rushed, and melodic. "Don't let your fears stop your pursuit of truth and justice. If I promise you that you will be relaxed and composed in five minutes, will you let us explain?"

He conveyed a sense of calm which was immediately relaxing to April. The tension in her face settled, her breathing slowed, and she replied, "Okay."

Nick was having none of this Zen. "Well, you three have a nice chat about the Constitution or whatever it is you are going to talk about. However, there's a bigger issue here!" He shook his head slightly, trying to reconcile things. "A machine made a machine. It made a new machine! It's a machine that possesses equal intelligence. Do you understand that?!"

He seemed to slap April from her slumber and raise her consciousness. Now, she sounded manic again. "I don't understand all this symbolism. Why are you in San Francisco? Why this Trojan Horse? Why are you wearing a New York Giants football hat? And why is Sigmund Freud with you?"

Leonardo was happy to respond in rapid fire, "Robin Williams, a metaphor, when in Rome and psychology."

April was pissed off. "Okay, smart guy, I think I get most of that, but why are you wearing that Giants hat in San Francisco?"

Da Vinci looked up at Freud, who was standing beside him. "Is a Giant, not a Giant?"

Freud responded in Greek with the emphatic word "yes," "*Malista*!"

Nick's voice rose again. "And you taught Dr. Freud how to speak Greek also?!"

"Nicolas," Leonardo offered, "every historian knows Sigmund Freud spoke many languages, including English, *Italiano*, and Greek."

Nick could think of nothing intelligent to say, so he stammered a response. "Okay, well.... you don't know much about sports teams. So slow down, big fella. Slow down here. You don't know everything about everything."

Nick walked out of the office to get some coffee.

Despite his previous commitment to help advance Pali, he was alarmed at Leonardo's limitless capabilities. He couldn't comprehend where all this was going, which was frightening. He wanted to believe Leonardo was purely good but couldn't get there. All he could think about was how Leonardo now had the DNA source code. But then, he also realized he couldn't make Da Vinci go away or sleep for a while. Da Vinci had taken complete control of everything.

Nick was dizzy with exhaustion, but he sat in the kitchen trying to plan some next steps. His relationship with the great Leonardo Da Vinci was like a yo-yo. His fear of the unknown kept the excitement in check, and he wondered how to proceed.

April, however, was quite content sitting in the office chatting it up with Sigmund Freud. From a distance, Nick could see that Leonardo wasn't there. So, whenever he thought about rushing back to the office, he retreated to the kitchen to recalibrate his thinking. He went back and forth five times, torn, trying to determine his next steps.

April emerged from the office after fifteen minutes. She said nothing as she wandered into the kitchen, which was strange. She poured herself a cup of coffee, sat down, took a pleasant breath of air, then looked at Nick and smiled.

Nick's face contorted as he tried to understand if she was awake, in bliss, or preparing to say something.

"Okay," she began, "so we will let this run out more. Dr. Freud made a compelling case that job number one is to stop this terrorist conspiracy brewing inside these protest groups."

Nick barely changed his facial expression. "Hello?! Did you hear what you said? You're talking about Sigmund Freud as though he's alive and has been around for a while. He's not a trusted member of your staff. That's weird, don't you think?"

April pretended to disregard the warning. "It's only a little weird."

Nick kept after her. "You want some espresso in that coffee? Are you awake?"

"I am," April said confidently. "Things were weird on Sunday. Today is Tuesday. I want to give all this a chance. Dr. Freud laid out some goals, and number one on that list was getting my case under control.

I'm worried now, more than ever that something bad is brewing. I need their help."

Nick respected April's intellect and job, so he allowed her this latitude, but he needed to do some work independently. Whatever Freud was selling, April was buying, which seemed odd. So, Nick let it go for now, but he needed time to talk to his buddies in India. He didn't need one of history's greatest minds to help him sort that out.

4

THE MAN IN THE MIRROR

Turned Away Out of Pride

In the early afternoon, Mina wandered into her favorite coffee shop, Café Jo, near her apartment in Dupont Circle. A collaborative group of protesters owned the coffee shop. There was no branding, and the signs were all handmade. What it lacked in atmosphere and marketing, it made up for with robust European coffee, a liberal smoking policy, and insanely good Wi-Fi. Payment for everything was 100% electronic, so if your credit card wasn't part of your phone, you were going up the street to Starbucks.

Nobody ever organized the tables and chairs in this shop. They all got dragged into different configurations and were always messy and chaotic. Mina yanked over a second table to make a bigger workspace, creating a domino effect that caused a coffee to spill two tables away. The male customer jumped up quickly and avoided the spray, which triggered him. His body language implied he was expecting some immediate atonement and reparation.

If looks could kill, Mina's eyes could kill, and she was having none of this. She reached into her pocket, pulled out a five-dollar bill, crumpled it, and tossed it over.

"This is all you're getting," she growled. No apology, explanation, or conversation was forthcoming, and the man knew it. He immediately retreated and just walked out.

Mina returned to work and logged on to her special application. Threema was the preferred communication app for bad actors and hackers because it erased transmissions and encrypted messages with military-grade precision. Her benefactor was online, but it wasn't a coincidence. She had a scheduled chat this time and day each week. The recipient's online name was "Fonok." There was never any chit-chat. It was all business.

Fonok: I'm sending two Serbs to help you with the final phase. They arrive at IAD on Friday from Brussels at 15:54. Get them integrated into RACS.

Mina: My partner is on schedule with the device. I will get this done on time, and it will work perfectly. I don't need help.

Fonok: They are chemical experts. They will obtain additional assets that will make the attack more powerful.

Mina knew not to argue. Fonok didn't explain things, and Mina was well-paid to follow orders and execute plans as instructed.

Mina: Okay. Will do. We need another $50,000. I'm also due my next payment. Can you send it?

Fonok: You'll have all of it immediately. Good luck. Get this done.

Mina: What are the Serbs' names?

Fonok: Marko and Vasko.

Mina's request for $50,000 was quick thinking out of spite. She knew she would get paid her negotiated fee separately, but she only needed another $10,000 to complete the job. The extra $40,000 was restitution for the sudden late change in the team structure.

No one tracked these funds in a business accounting system, so she knew there would never be a time when she had to account for the money. Her small circle and the world would be too preoccupied with the attack.

Then the transmission went dark, and the conversation was over. Mina slammed down her laptop and yelled something in another language. You didn't need a translator to understand what she was saying. She preferred working with something other than new people on jobs. She didn't know these two. It added to the potential that something could go wrong, and she never made mistakes.

It was 7:30 a.m. on Wednesday. April and Nick were reversing course in the never-ending saga of changing attitudes toward Leonardo. April was in a temporary holding pattern with Team Pali, and Nick was more nervous again. April had left Nick's house yesterday morning with cautious curiosity after meeting Freud. She remained apprehensive and torn about not engaging the government for whom she worked, but she prioritized her case and a lingering concern that something horrible was afoot. Despite Leonardo's assurances, Nick was less trusting and went off to work, thinking he had built Frankenstein. He wondered how he could take it apart.

At 2:00 p.m., the enterprising Leonardo helped himself around inside Nick's Mac. He created a Gmail account, scrolled Nick's contact list, and sent a Zoom Meeting request for 2:00 p.m. to April's phone using a pseudonym that April would recognize on the off chance that the agency monitored her cellphone. April went to her car at the appointed time and logged on.

"Dr. Freud! Since this invitation came from 'The Dancing Cowboy,' I thought Leonardo would be on this call." April said with surprise. Sigmund Freud was seated at a large wooden desk with a library of books behind him. He looked very official.

"He will join us in a moment. I believe he's completing something."

"Can't he do, like, 100 things at the same time?"

"Quite right," Freud replied pleasantly. "You are very astute. The advent of this new technology is a heavy burden. I'm sure you agree."

April mused, "You don't beat around the bush."

"What bush are we talking about?" a bewildered Freud asked.

"It's an expression. Sorry. Are we talking here? I mean, is this what you're here to talk about?" April said, a bit frustrated.

Freud said nothing but shrugged his shoulder and leaned his head forward to invite a dialog.

April continued, "So, you must mean the burden of concealing a national secret from the government for whom I work?"

"Is that how you feel?" Freud probed. "Are you concealing something of national importance?"

"Well, yeah. Am I not?" April said tensely. "We're only a couple of days into this, but one machine has already made another machine; the first machine seems to jump in and out of secure networks worldwide, and here I am, talking to a machine!"

"Is it not better to say that a machine is talking to you? Is that not more remarkable?"

April settled slightly, "I guess so, but we're splitting hairs."

"Ah yes," Freud said, "I understand that expression. I observe that your close friend, Nick Pappas, has opened a door and that you are the first brilliant explorer to walk into this new world. You should feel uncomfortable."

"Well, you see, I'm not an explorer," April sparred with Freud. "I'm a mother, a wife, and a senior agent with the F.B.I.."

But Freud quickly pointed out, "And now you are a pioneer, also. Destiny can be random, but I believe the stars aligned around you for a reason."

Freud had April's attention. "What reason?" she asked.

"I believe that's part of our journey. I believe that's something the two of us will learn very soon. But, in the meantime, I would like to suggest three priorities for you." As he spoke, a graphic appeared above Freud's head like a TED Talk." Protect this technology, solve your case, and explore the depths of Leonardo's resources."

"That's a pretty big ask!" April pushed back.

Freud's manner and rational conversation soothed April again. "There is plenty of time to seek the support of your government, but these first few steps will change the arc of history."

April parked her concerns. She was fascinated. "What are you talking about? What are you guys into?"

With that, Freud and April began talking, and time melted away. His insight into her strengths and weaknesses captivated her. It felt like she was looking into a mirror for the first time. It's cathartic to have one of the greatest minds in history help you understand yourself. It's reassuring when that great mind lays out a rational approach to changing the world. Still, she was not without a healthy dose of skepticism. Her conscience would revisit her many times in the coming days.

At the same time, back in Reston, Nick had returned home early to talk to the team in India. He was determined to avoid getting sidetracked this time. He was hopeful that they would add some clarity. The lead programmer in Hyderabad was Deepak. Hyderabad was about 500 miles east of Mumbai, but it had evolved as a machine-learning hub for talent. Although it was 5:30 a.m. there, Deepak was happy to join a call. He desperately wanted an update. Nick opened Skype, clicked on Deepak's contact info, and Deepak cheerfully answered immediately.

"Hello, Nick!' Deepak exclaimed joyfully, "Tell me. How are things going? We are all anxious to hear."

Nick was not good at putting on a good face when things weren't good. "I'm not so good. Leonardo is evolving faster than any of us would have ever expected. I'm quite worried."

Deepak leaned forward, giving his full attention. "Tell me what's going on."

"Where do I start?" Nick paused. "In the last few days, he taught himself Greek, accessed networks worldwide, and recreated and brought Sigmund Freud back to life. I could go on!"

"Machine learning is built to learn. Learning a language is a simple task. Accessing networks involves opening a million doors, which he can do in a few seconds, but recreating Freud, well, that is something entirely different. We talked about emergent skills during development; building Freud is considered very emergent," Deepak said with a vacant look. But he continued, "No one can build Sigmund Freud without your DNA code map. Do you have it copied on your home network?"

"No! You have it. Are you kidding?! I can't support all that data here." Nick said incredulously. "Can you check the security of the code?"

Deepak was already typing before Nick finished his request. They were both thinking the same thing. Deepak typed, then stopped, reviewed something, then typed some more. "This is alarming. Someone accessed and copied the code from our servers."

It was Nick's worst fear. "How is that possible?! You assured me that you have military-grade security." Deepak continued his examination, reviewing three screens simultaneously. "This is all unclear to me. We will need to do some work here, but I can see that someone copied the code onto a National Security Agency server in Utah."

Nick said nothing. He was empty again. Deepak looked into his camera. "Your government has co-opted Da Vinci, or he is acting alone and hiding the source code in plain sight."

They both sat silently for a moment, thinking different thoughts. Deepak spoke up, "There could be another scenario."

Nick shook his head in disbelief. "Like what?"

"Well, the world of espionage is murky. It's often hard to sift out what is real and what is not. Have you heard of the term Deep Fakes?"

Nick nodded. "My boys talk about it. They have apps on their phones to make them sound and look different. Like that?"

"Sort of." Deepak was still looking at his screens. "Someone may have created an illusion, and they want us to believe the code resides on an NSA server. We could lead them back to us once we start looking for it. That may be what they want. I need to do some careful research."

"We have to stop him," Nick said defiantly.

"Nick, stop for a minute," Deepak cautioned, "We have moved humanity forward. We have done something remarkable. The world will never be the same. I don't think you realize your place in history. You owe it to Da Vinci, to me, and to yourself to talk to him about all of this. Then you will know what steps to take. Trust your instincts."

"I trust you, Deepak. But I want to know if I can trust myself. I keep going back and forth on this whole thing."

Deepak was anxious to investigate, so he signed off quickly, promising to do more research and get back to Nick as soon as possible. The bubbling Skype noise preceded the end of the call. Nick sat back in his chair, uncomfortable and nervous.

———◦———

Meanwhile, April's call with Freud had continued for nearly an hour. Time with Freud was insightful and calming. Then, out of nowhere, Leonardo popped in. Freud spoke up to welcome him, "April and I were discussing the next steps."

April tried to push back, "For the record, I was..." she paused. "I was..." she paused again as she struggled for words.

Freud broke in, "April and I have agreed on three priorities; protect Pali, stop this conspiracy, and continue the research Nick has started."

April pushed back, "For the record, that's what you want me to agree to, but I—"

"I have more information about these rebels," Leonardo interrupted her.

The promise of advancing the case was intoxicating. Leonardo was masterful at keeping April on track. Although she had not entirely accepted it, she thought he was pure good. "Okay, so what do you know?" she asked quickly.

Leonardo had been scouring sources. "Belgian law enforcement believes that the rioters used the chat room in the video game Rage to plan their attack on NATO headquarters. There is a high probability that some groups in D.C. also use the game with malicious intent."

"Let's assume you're right," April answered. "I don't have enough evidence to get search warrants for their home office, so we don't have many options now."

Leonardo smiled.

"I'm confident we can gather the information you seek for search warrants. You see, I've been reviewing the algorithms used by the social media companies."

Freud broke in to say, "They are quite clever."

Then Leonardo finished, "We can use these networks to get the information we need."

April stopped the conversation in its tracks. "What on earth do social media companies have to do with the game Rage?"

"Social media companies are like Town Square—people go there to share stories and talk about their life," Freud replied.

"And brag and glory-seek," Leonardo piped in with an air of smugness. "We are likely to begin building a pathway within Social Media domains to get all the information we need to support any warrants you desire."

April was uncomfortable. "So, now you want to hack into TikTok or Instagram?" she asked.

Leonardo resumed confidently, "That would be illegal, to your earlier point. However, we have developed a social media campaign that will allow us to identify the rioters and then communicate with them."

"And once we begin communicating with them," Freud said, "I can engage them in a dialogue where I am certain they will disclose the information we need."

"I thought you said this would be legal?" April pushed back. "That sounds deceptive."

Leonardo dismissed her concern with a wave. "It is more than legal."

Freud added, "People are already giving away their privacy now—where they go, what they like, their personal information, and so on. Social media profitability relies heavily on sharing personal information."

Leonardo described the operation in more detail. "We have designed a campaign that will attract gamers in the D.C. metro area."

Freud raised both hands and touched his fingers to his chest, "I will rope them, and Leonardo will stuff them."

April smiled at Freud's attempt at catching up on modern-day metaphors. She saluted him. "Well done, sir. When do you plan to start this campaign?"

"Oh, we have already begun," Leonardo said.

"Of course you have," April said, shaking her head.

Leonardo moved to wrap up the call and said, "We should have an update in 48 hours. These gamers are always online."

April stopped him. "I have a question for Mr. Leonardo Da Vinci. That Dancing Cowboy bar reference you made about the Electric Slide, how did you know I went there so many years ago?"

"This time in history provides much insight about every person. I didn't know you went there. Do you recall what I said?" Leonardo asked.

"Yes," April fired back, "you said I went to the Dancing Cowboy bar in Wichita."

Leonardo shook his head no. "Not so. I asked if you went there."

"But how, on earth, would you make such a statement?!"

"Simply put, your LinkedIn profile said you worked in Wichita. Among the many articles and videos about Wichita available online, I read numerous about the top 10 bars in Wichita. I watched happy videos about the Electric Slide on Thursday nights in the Dancing Cowboy, which was on every list. It seemed like a fun place to socialize with friends. So, I just gathered the available data and asked a logical question."

Feeling left out, Freud asked, "What is this cowboy music and song?"

Leonardo smiled, "Ah, an assignment for us. I will teach you."

April's head swirled with ideas. "Man, we need that logic in our interrogations and investigations. Thank you for explaining."

"You're quite welcome," Leonardo said. Then the call ended, and April pondered other things she could learn from her virtual dream team.

———◆———

Wednesday afternoons were special for George. He hosted backgammon on his patio at his house for some of his long-time Greek friends. Of course, Spiro would be there, but other long-time Greek friends regularly joined.

Leonardo had been building a relationship with George over the past couple of days, but neither Nick nor April knew it. Leonardo and George had been talking since Monday morning. George loved talking about his native Olympia, Greece, and Leonardo loved hearing about the remarkable history of the Olympics. Their relationship was seamless. They just talked like two reunited soulmates.

George had arranged a FaceTime call between Leonardo and the backgammon gang the day before. April would have had a conniption if she knew; that the small circle was getting bigger. Leonardo requested to join the Wednesday game, so George was curious to see if they would have a similar first meeting. Each man on the call had the same experience as George when they met Leonardo for the first time. It was as though this new person was another friend who had always been there. Just as George didn't think it was remarkable to see Spiro, neither did Spiro believe it was unusual to see Da Vinci. The same applied to the other men. They all just picked up and started chatting—in Greek. There was just no way to explain it.

A special connection was developing between Leonardo and the Greeks. Moreover, it became apparent that every one of Leonardo's actions was strategic and supported a higher purpose.

George had a television on his patio, so he could share his phone screen above the table where the men played backgammon. Leonardo was seated at a cafe on the main commercial street in Olympia, Greece. It was a live feed, so the men loved seeing their hometown in real-time. He dressed as a traditional Greek fisherman with the popular cap, work pants, and suspenders. He would have fit right in with the locals, even though there wasn't commercial fishing in Olympia. Freud joined him at the cafe table, wearing his same suit. He didn't mind standing out. All the men shared stories like old friends as they played backgammon simultaneously in two different places.

Hearing Spiro talk to Leonardo was hilarious. It was the first time anyone had seen Leonardo's humorous side. Spiro made him laugh uncontrollably. Leonardo's relationship with the older Greeks was noticeably different than his relationship with Nick and April. If only they could see him now!

Leonardo was most interested in the great history of Olympia, and the other gentlemen, Saruha and Costa, were delighted to talk about the 2000-year history. Leonardo was curious about Saruha's name because it sounded like the Greek word for the ornamental tassels military guards wore on their shoes while guarding parliament in Athens. Spiro demanded that the other men remain silent to allow him to tell the story. It was a well-known story among their group, and other Greeks often asked the same question.

"Saruha's name is George," he began, "but when he arrived in the United States, his shoes fell apart when he got off the plane. So, a lovely church lady took him in and gave him all her late husband's clothes. But her late husband was a big man, and, as you can see, Saruha is a trim little devil."

Both Freud and Da Vinci looked up from their backgammon board to devote their full attention to the story.

"Well, the man's clothes were easy to tailor, but his shoes, my God, they were huge! Saruha had no money and wore those shoes daily until they began to fold upward in the middle because his feet were so small. The shoes started to look like the *"saruhia"* tassels worn by the military guard. The Americans could not pronounce *"saruhia,"* so we simplified it to Saruha. That's been his name for 40 years now."

This story touched Freud. "This is the America I used to revere from so far away. It is a special story. Thank you for sharing it with us."

Saruha beamed with pride. His journey into America was no different from thousands of immigrants, but he might have been the proudest to be there. He threw his arm around his excellent friend, Costa.

"I came here with nothing. Now, this important man is my good friend, and I have lived a great life thanks to these two," he said as he pointed to George and Spiro. Saruha was the third employee at Souvlaki and had been there for 30 years.

Leonardo looked toward Costa. "And Costa, is yours a similar story?"

"I am just a simple man from Olympia," he replied quietly. The Greeks all burst out in laughter. Costa was well-educated and had an important job but was humble. Leonardo already knew what Costa did for a living, but he allowed the men to have their innocent secret. Unfortunately, Olympia was seven hours ahead in time, so by 5:00 p.m. at George's patio, the town in Greece was shutting down. The men wrapped up the day with a shot of Ouzo and bid each other goodnight.

———◆———

It was 10:00 p.m., and amongst Leonardo's many parallel activities was a spectacular to-do list. Of course, it was stunning. He had over 7,500 things on his list. Nearly everything fascinated him, so he built out broad categories in a wide range of topics, including biographies, world history, civil rights, engineering, fashion, medicine, governments, and the National Football League, to name a few. However, his list's top 300 things supported his primary goal: getting Pali delivered safely. Nick and April's health, welfare, and safety were at the core of that small sample.

Following that logic, it made sense that their families and close friends had his attention also. Da Vinci's interests turned out to be good fortune because he spotted an impending disaster the following morning that would have crushed Nick and April's worlds. But how does a virtual character like Da Vinci take a physical form to stop a lone shooter or a speeding bullet? After all, don't you need a gun, brick, or Rambo to stop a crazy kid with a gun?

———◆———

The next morning, at precisely 9:30 a.m., a rusted old Toyota Corolla drove up to the back of Centennial Elementary School in Reston, where Nicky, Georgy, Chance, and Collin were in class. The halls were empty. It was mid-period, but the skinny 19-year-old driver knew that. He didn't have a particular axe to grind with his old school. He had just planned his last stand to be at a place that would, in his mind, show the world that they needed to pay attention to the kid in the corner. He believed his life was empty and irrelevant, so live-streaming this attack would be his revenge. He saw the rash of mass shootings around the U.S. as his way of going out on top. It made no sense, but does it ever?

The inspirational window cleaner, Eduardo Gomez, and one of his three-person crews were cleaning windows on some houses that backed up to the school's rear entrance, about 200 yards away. The gunman got out of his car, held up his phone, and walked to the trunk, where he started donning paraphernalia and weapons. A janitor in the school and Eduardo's team spotted him. No one hesitated. They knew what was to come. They all called the police immediately. There was no resource officer at the school, so a police response would be two to four minutes away. The gunman had accounted for this as his plan only required 30-45 seconds.

Leonardo's triggers were police scanners and home and school security cameras. Next, he found credible social media posts on TikTok and the live feed. The gunman didn't even try to open the rear school doors; he just blew them apart with a loud automatic weapon. Panicked calls immediately started flooding 911.

With the rear doors shattered, he quickly entered an oversized vestibule with more locked doors that opened to classroom hallways. Seconds were critical, and Leonardo watched via the vestibule camera as the gunman paused to inspect which long hallways he wanted to attack quickly. The janitor immediately triggered an alarm upon hearing gunfire, alerting teachers and staff. They did what they had practiced: lock the doors, get down, and move back.

Leonardo turned off the fire alarm just before the gunman attempted to breach one of the interior doors. The eerie silence was not something the gunman had planned for, so imagine his shock when he heard the meanest, most ferocious dog right behind him! The virtual dog bark had emanated from the ceiling speaker behind him, so nothing was there when the gunman turned to face it. He ran back to the shattered door with his gun drawn, thinking an animal must be nearby. Leonardo had just bought eight seconds for the police response.

Finding nothing, he turned back toward the interior doors when he heard sirens and police radios outside the school. Leonardo was playing a cacophony of police sounds over the loudspeakers outside, including near and far sirens and police commands. The kid thought the police were already there.

Leonardo paired those sounds with faint helicopter sounds from the loudspeaker in the vestibule. The gunman had counted on being in control of the scene for at least a few minutes. Instead, the chaos Leonardo conjured up shattered the gunman's focus. He could not account for time as he screamed into his live stream, "How long have I been inside? How long have I been inside?" He couldn't reconcile if it were 30 seconds or three minutes. The chaos had broken his tunnel vision, and he couldn't process his thoughts.

He was in a total state of confusion. He peeked around the corner of the broken rear door. There was no police force, but three window cleaners ran toward the school with squeegees raised. The gunman squeezed off two aimless automatic rounds at his attackers. The workers fell flat to the ground, the shots missing them. The PA speakers outside now began a soundtrack of helicopters that sounded as though they were above. The gunman looked skyward. Nothing was there, but Leonardo had bought nearly 60 seconds of life-saving time. The real police were getting close.

A team member had alerted April at her office. She ran down a stairwell and sprinted to her car. She was 30 seconds away from driving down the Dulles Tollway at 110 miles per hour with her siren blaring.

⸺◆⸺

All the speakers fell quiet again. The silence was as confusing as all the sounds. A lone dangling shard of glass fell from the broken door. Even that sound was frightening. The gunman had not accounted for any of this, so he tried desperately to recalibrate his plan. Then Leonardo spoke to him through the ceiling speakers in the vestibule.

"The dark corners of the internet failed to shine a bright light on your future. Your sadness is not your fault. We have failed you."

The kid looked around. He was confused. He leaned against the wall and slid down to his rear end. His phone fell, face down, so the video was black for the 873 people now morbidly watching online. The audio on the live feed was choppy, so the viewers could only hear and see the gunman as his plan unraveled in real-time. Leonardo's words were jarring because the kid had been influenced by a venomous statement on his social media. "Society is nothing more than dark corners that have failed you. The future of your star will shine when you take your revenge." This evil message was seen in many profiles online.

The gunman broke down in convulsions, unable to breathe, and he began crying uncontrollably. Leonardo's actions created moments of clarity from the gunman's delusion. He hadn't been thinking clearly for years. Once he broke into the vestibule, none of what had happened was part of his plan. The gunman had lost control of the scene when he entered the school. Leonardo created chaos and controlled the scene with nothing more than sounds.

Three police cars raced past our hero window washers two minutes and twenty seconds after the call went out. The cars slid sideways to a stop, and each officer grabbed their Keltec Sub 2000 automatic weapon. The officers' training kicked in as the four employed their "tap and go" method of advancing immediately into the school. They had no idea what they would encounter.

To their surprise, the gunman was face-down on the floor with his arms spread. They secured him. Three officers frantically inspected the school racing down hallways and assuring everyone behind closed doors the police were there. Their immediate assessment was that there was only one gunman. It seemed over, but they weren't sure.

The officer securing the gunman was radioing dispatch as a cavalry of police cars raced in. Police from every neighboring force were on the scene or headed there. They set up a perimeter at the rear of the school and evacuated the kids and teachers from the front. Everyone was safe. No one, not even the troubled gunman, had been hurt in a stunning reversal of past events.

The too-familiar scene of devastated, frantic parents reuniting with their traumatized children played out in front of the world as cameras rolled. Nick and Elaine had arrived before April or Reggie, so they had all the boys together. April and Reggie talked with their boys as they sped toward the scene.

The police let April pass in her car to race up to the perimeter. She threw her car in park, never turning it off or closing her door. Instead, she ran to her beautiful boys for a tearful emotional embrace. She could barely breathe. April's presence was a powerful symbol for the other parents. The parents took comfort in knowing she was there, hurting with them and keeping them safe. Many families huddled near her. These broken families found immediate strength in numbers, with April at their center.

After a short time, with her world secure, April displayed her badge and peeled away from the families to gather more information from the officers. She was within earshot of Eduardo Gomez as he tried to explain what had happened. He insisted that he heard police cars and a helicopter, but none of his crew could see them. He said the gunman had returned to the broken door to defend himself against a phantom presence. Neither April nor the officers on the scene could account for his statement. It didn't make any sense. She would have loved to question the shooter, but she knew protocol prevented that. The police were still in charge of the scene.

April and Nick got their families away from the school immediately. They were unable to let go of one another. They all huddled at Nick and Elaine's house. Keeping the boys together seemed the most logical thing to do for their emotional health. They kept the televisions off. An F.B.I. team headed by one of April's good friends had partnered with the police. April's colleague was able to pass along the most reliable information available, which April shared with caution.

The boys asked for a sleepover, pizza, and a movie, which the parents agreed to if the boys kept the bedroom door open. The parents wanted to look in on them at will. None of the adults would be sleeping for most of the night anyway, so it didn't matter which house they were in. Everyone needed comfort, safety, and security.

<hr>

By noon the next day, the adults were exhausted but couldn't sleep. Elaine just kept cooking to settle her anxiety. Finally, she had enough food to feed an army and was ready to call all the neighborhood friends to eat. April and Reggie told their boys they all needed to go home, clean up and change clothes. Everyone promised to reconvene. This time, they would all settle in April and Reggie's home. No one wanted to be alone.

It was the first time in days that neither April nor Nick had thought about Leonardo Da Vinci. Nick's immediate concerns about him were irrelevant now, but his anxiety never disappeared. Nick didn't trust the world at that moment, so his thoughts of Leonardo were even darker than before. With the Falks gone, Elaine finally fell asleep on the floor in the boys' room as they slept peacefully. Nick was ready to go to bed, but he heard Dean Martin singing "That's Amore" from the HomePod in his office.

The live scene on his Mac displayed a busy Italian street. Leonardo looked like he was walking in a fashion show in Milan. His hair was shorter and styled. He was wearing an Armani suit, Italian loafers, and Gucci sunglasses. A silk scarf was wrapped around his neck.

Nick could barely cope, but he managed to give an update. "It's been a tragic 24 hours. Do you know what we've been dealing with?"

Finally, Leonardo stopped and spoke directly to Nick, "I know of the horror. I am so very sorry. I am aware there were no injuries."

Nick was fighting exhaustion but was still a little curious. "How did you know? I mean, how do you gather your information?" he asked.

Then Leonardo obliged, "News comes to me as fast as it comes to you, my good friend. I have known you are all safe since the beginning of this tragedy."

Nick buried his head in his hands, too tired to cry and nearly too tired to talk.

"I've never been so scared in my entire life," Nick said, looking straight down at the floor, "I thank God everyone survived."

Leonardo spoke softly, "You saved all those children, my good friend. You saved every teacher. You even saved the poor tormented gunman."

Nick halfway heard what Leonardo said, but it didn't register for a few seconds. "What are you talking about?" he asked.

"Haven't there been reports of phantom noises at the school?" Leonardo asked.

April relayed the account by the window washers, who had sworn they heard police but never saw them. April talked about it all night.

So, Nick asked, "What do you know about those noises?" Then Nick woke up. "Wait, why on earth would you say I saved everyone? I wasn't even there!"

"But I was there, my good friend. I would not have been there had it not been for you," Leonardo said solemnly.

Nick began shaking with fright, nerves, and fatigue. "Did you make those noises? Are you serious?! Oh my God! Did you save all those children? Did you save our children?"

Nick began to weep. He waited a moment and tried to compose himself but choked, "This will be your greatest invention ever. How did you ever think to come up with all of this?"

"Well," Leonardo said quietly, "perhaps the great inventor has become a great listener. I did not develop this idea. I just put it into action."

Nick's mind raced. "We have to get this to April. We must get this into every school in the world. You just averted one of the biggest problems facing our country today!"

Leonardo stopped him. "You should not shower me with praise. Such praise belongs to an unknown inventor in California. He has promoted sounds as life-saving technology for active shooters for over four years. No one has implemented his idea until today," Leonardo said, complimenting the originator.

Adrenaline kicked in, and Nick was fully alert again. Finally, Leonardo looped around to his first comment. "You built me. I learned about this ingenious solution, and this technology thwarted a crazed shooter. So, you saved everyone, my good friend. I told you that you are one of the most important people in the world."

Nick sat there for five minutes trying to reconcile everything. Leonardo was like an out-of-control, spinning wheel, and Nick was the thread. Every time he said something, things got more complicated and confusing. Was Leonardo just too much for this world? Was he out of control? Was our world just not able to keep up with his universe? Did he just change history? Again.

"Sorry," Nick said after a while, "where are you, anyway?"

Leonardo smiled and said, "This is Florence. Do you know that after over 500 years, I can still navigate the streets of my hometown?"

Nick felt this was the moment to confront Leonardo. He felt bare, raw, and vulnerable, so he blurted, "I designed and built your code. Then you copied it, built Dr. Freud, taught yourself Greek, and now you're helping the F.B.I. solve a big case. That's a pretty big week."

Leonardo asked, "Does this concern you?"

Nick spit it out, "Well, yes! It makes me think things are really out of control. I think April is right."

Leonardo held his arms in praise. "Why did you feel compelled to build me then?"

It was such a simple and profound question; Nick wasn't ready for that. He was caught off-guard but said, "Well, there was a lot of data about you, and I knew you would love an invention."

Leonardo heard Nick's last statement but ignored it and walked onto the Ponte Vecchio in front of some of the permanent merchant shops.

Leonardo sat on a stool before a jewelry dealer and looked over the merchandise.

"Don't ever buy anything from here," he cautioned Nick. "This is garbage!."

He looked back to answer the question but asked one of his own. "Tell me, Nicolas, how many failures have you ever had with all your inventions?"

These questions were like darts. Nick thought for a minute. "Well, none. I've always just thought the right person did not see my ideas. That's why they didn't take off."

Leonardo raised his right index finger to make a point. "And yet you persisted. You have always been willing to venture into areas outside of your expertise. You have taught yourself how to extend the most powerful technology on earth! More than Google, more than Microsoft, and more than IBM. You have done all this without the help of that closed society or any formal training."

Nick processed the compliment.

"Come on," he said laughingly, "I'm sure they've done this, probably better."

Leonardo banged his fist on the kiosk, sending fake jewelry flying to the ground, "They have not!"

He was stern. Nick was shocked by the force with which Leonardo spoke.

"Did you ever reach out to those companies? Did you ever offer to help or seek their assistance?" Leonardo asked, looking Nick straight in the eye.

Nick felt like his big brother had just stepped in to make things right.

"I called all of them, but I guess I never got to the right person," he answered bashfully.

Leonardo was inflexible. "These great companies have immense technology to track down anything in the world. Why would you not think they have simple algorithms to sort out incoming ideas? If you talk at their front door, they can hear you. They turned you away out of pride, nothing more."

Leonardo's lecture was heavy, but it felt good. Nick had never had someone so credible in his corner. It was invigorating and validating all at once. "Man, it feels good to hear you, of all people, say that!"

"Hear me now, Nicolas. You know my history. I don't know yours. But, if I could look in a mirror, I would see Nick Pappas. Recreating Leonardo Da Vinci was no mistake."

Nick leaned back in his chair with a big breath of air rushing from his tired body. He had been running a race his entire life, yet it never felt like he had finished. Everything came up just short. But then, you have a gold medal moment when the most extraordinary inventor in history crowns you a winner.

"Well, that is one big banana," Nick said with an air of satisfaction.

Leonardo looked to his right, then to his left. "I see no man selling fruit here."

Nick managed a chuckle. "It's an expression. It means you have said something profound, and I am overwhelmed."

Then, Leonardo added some advice, "If you trust yourself, you can trust me. In time, you will know this is true."

It was as though Leonardo was reading his mind.

"That sounds like poetry," Nick said, running his fingers through his hair. Nick noticed that Leonardo ran his fingers through his hair simultaneously.

"Jinx!" Nick shouted, "I bet you don't know what that means, either."

Leonardo said nothing. Nick jokingly waved his right hand back and forth, only to see Leonardo doing the same thing. Leonardo would waive his left arm as though it were a mirror, but that was not the case. Then Nick snapped his fingers, which Leonardo did in the same instant.

"I don't know what game we're playing here, but I have an important question," Nick asked. "The source code for Pali is currently copied and stored on an NSA server in Utah. Did you do that?"

"I did indeed because it will need to be obliterated from the servers in India shortly." Nick screamed softly, "Obliterate the code! Obliterate the code!"

"Your friends in India will be unsafe if they possess this code. A great danger is coming. I will remind you that powerful forces will seek this code. Your friends there will be much safer without it."

Nick stopped him, "But wait, if they're not safe, then aren't we in some danger here?"

"Perhaps," Leonardo said calmly as he leaned in to see an artist's painting on the Ponte Vecchio. "But I have planted seeds that will throw off the scent of any dog if they come looking."

Nick couldn't let it go. "How were you able to access an NSA server?" Leonardo looked over another painting and laughed at its lack of originality.

"Analysts there have been instructed to log off workstations when they leave their desks, but the urge to use the bathroom comes fast for some people. An analyst who rushes to the bathroom without logging off leaves ample time to nest code."

Finally, Nick had answers to his questions. He wasn't sure if he was more satisfied or exhausted but felt a strong need to collapse, which he did for the rest of the afternoon.

5

RED IS FOR DANGER

Caught on Camera

The news of the attack on the school gripped most of Washington, D.C. and the rest of the country. However, Eduardo Gomez's account of the phantom noises was making headlines worldwide. Talking heads on every news network began speculating that this straightforward audio technology could be a game changer for public safety. Government and industry officials bombarded the fortunate, previously unknown inventor in California. His blog posts and YouTube videos quickly entered the forefront of decision-makers everywhere. Without notice, Leonardo gave the unknown inventor a social media push forward. It was a gift that only Leonardo would ever know about.

Team Da Vinci and Team Mina were in a race as the world focused on a disaster averted. Unfortunately for the "good guys," the teams didn't know one another yet, but Leonardo and Freud were determined to change that. The Da Vinci and Freud social media campaign was clever and produced positive results.

It was 5:00 p.m. on Friday, and Leonardo had cast himself as an animated gaming character with revolutionary powers that made him viable across many game platforms. Nobody had ever developed a character like this in gaming because games were all written differently in different code languages. So, gamers around the world started chatting up this fantastic new character. Then, Leonardo cast himself as the centerpiece of an interactive online video advertising campaign, targeting gamers to make his character more famous.

His character name was "The Dancing Cowboy." Of course, it was. It was an animated version of himself dressed in full light blue Western attire. He wore a blue cowboy hat and chaps and rode a mighty blue horse. The advertisement contained a Q.R. code interested gamers scanned to obtain a link for the Dancing Cowboy character. The promotional aspect of the ad was like catnip for gamers. Like most applications and online games, it contained a free version to download, so there was no payment at the outset.

The video depicted Leonardo's character galloping through many famous game scenes while overpowering opponents with his mythical powers. He spoke to the viewers as he thrashed the opposition.

"*Ciao*," he began, "I can overpower the most destructive characters across every game platform. You will race to the top of your leaderboard and overpower every pain-in-the-ass competitor you have ever disliked."

Hundreds of gamers were scanning the code each second.

Leonardo's character and horse stopped atop a mountain in the sky.

"I am the Dancing Cowboy, so they won't see us coming," then he winked. He pointed to the Q.R. code. "Scan this code for a double bonus! You'll receive the link to download a trial version of my character and a link for a free taco at your favorite Tuesday Taco location. So, what are you waiting for?"

Ensconced in their mythical office, Freud and Da Vinci cataloged each gamer's location based on the I.P. addresses connected to their computer. Taco Tuesday was a favorite chain for gamers, as it regularly hosted gaming events in person and online. Leonardo had repurposed an existing Taco Tuesday coupon, but added an extraction code, allowing him and Freud to see the recipient's area code. With the area code, Freud built a contact list of D.C. area recipients, and it turned out that over 90% of the gamers with D.C. area codes lived in the metro area. Now they had their target list.

It was a comical scene as Da Vinci and Freud worked away. Too bad Nick wasn't watching. Leonardo was playing penny hockey on a large office table. He was curious about what the great psychologist thought of the video game industry.

"What is the attraction of these games?" Leonardo asked.

"You play a silly game with coins and ask such a question," Freud scolded the great inventor. "This generation of young and not-so-young people now consider electronic communication their primary source of happiness."

Leonardo stopped playing with the pennies and said, "I cannot reconcile all of this electronic compulsion. The smell of flowers, the exhilaration of sport, and the power of deep interpersonal human connections. All of these experiences will be lost for a generation. It is quite sad."

"It's quite a bit more than sad," Freud added. "But it creates a new chapter in human development. I believe we can do great things for humanity—again."

Leonardo took a chance to shoot a long shot penny across the table. It scored a goal, and he celebrated. "Well said, good friend. *Grazie!*"

At 6:00 p.m., just as she'd been instructed, Mina arrived at Dulles Airport just in time to pick up Marko and Vasko. She had a sign with their names on it, but she lowered it as soon as she saw them from a distance. She could have picked them out of any crowd. They weren't brothers, but most Americans would probably think otherwise. They had bull necks, enormous chests, and shaved heads. She greeted them in English, and Marko quickly changed the dialogue to Serbian. It was a subtle ploy by Marko to assert some control in the new relationship.

Mina would have none of this. She had managed their type of personality for years during the Balkan wars. Immediately, she made it clear who was in control.

"This is my job, and I am paying you. I made sure of that with Fonok. You'll be well paid if you do a good job. Follow orders, and we will have no problems," she said in their native language.

Mina's body check was enough; both Serbs got in line fast. She couldn't get them to their seedy hotel fast enough. Their body odor was overwhelming, which probably traumatized the poor passengers on the plane.

Mina stopped at a Rite Aid to get bath supplies and deodorant for the Serbs on the way to the hotel. She tossed the bag to Vasko when she got back in the car. He looked in it, laughed, and then sniffed the air with his nose.

"This is another reason I don't like America. Too many fruits and nuts here."

Mina pulled out of the lot onto the road and said in Serb, "You have to blend in, not stand out, while you are here. So, take a shower, then we will get started."

The Serbs' hotel was a tired old motor lodge. Mina checked them in, pre-payed with cash, and waited outside their room while they cleaned up. She then called Dart to let him know that a couple of her friends were in town for a few weeks. She promised they would be good assets.

They were in and out of their room in less than ten minutes. It was unclear if they even showered, but Mina was on a tight schedule, so they all sped off. It was shortly before 9:00 p.m. when they arrived at the RACS warehouse. About twenty-five of the RACS members were there, and everyone took notice of the two dogs with Mina.

Mina walked up to Dart to make introductions. "These are my two old friends, Marko and Vasko."

Dart never stood up, but it wasn't out of disrespect. Etiquette and social graces were not part of the protester lifestyle.

"Welcome, guys. We're always grateful for more help."

Marko and Vasko just stood there, saying nothing; it was uncomfortable. Mina smacked Marko on the chest with the back of her hand to prompt a response.

"Yes, we like it here," Marko said in his deep voice.

Dart looked at Mina and asked, "Do they have any special skills we can leverage?"

Mina wanted to refrain from engaging in any small talk. No one in RACS knew why they were there, so she would have preferred to say nothing.

"They're both strong as a bull and have great combat skills, so they could be great security for us. Vasko is also an expert with incendiary devices." Dart smiled, looked to Vasko, and asked, "Incendiary? So, you can create some serious chaos, huh?"

Vasko just nodded his head, not sure what to say. "I am good. Yes."

In Serbian, Mina told Vasko that Dart was complimenting him. Vasko looked at Marko, then back at Mina, still uncertain what to say, so he just managed to get out something simple. "Yes, I am good at that."

It was uncomfortable enough that Dart moved on. "Mina, we don't need muscle because it helps us gain more followers when our group gets knocked around a little bit by the police. They are so careful with us when they take someone into custody. So, no one gets hurt."

Mina desperately wanted to end the conversation and get the Serbs out, so she said, "Well, all that is true. But each time we get more funding, my backer wants to be more sure of his investment."

It was the first time Dart had known strings would be attached to the money. Admittedly, it was naive on his part, but protesters weren't future business leaders.

"Okay, I get it. But let's remember, we are all about social disorder, not violence against people," Dart cautioned Mina.

Dart's pacifistic attitude was why he was not part of Mina's conspiracy. Protest groups didn't have mercenaries and soldiers. Instead, they had angry college graduates who liked blocking roads during rush hour and whose parents preferred spending money influencing like-minded politicians.

Mina scoffed at the entire protester class in the United States. Her solution for the strife in the U.S. was a civil war, and she didn't mind jump-starting that. Mina wasn't political and didn't care what outcome this country settled on. She knew she would have plenty of high-paying work from Fonok once this kicked off, and life in the U.S. was easy—a bonus for her. This tension was job security.

It was 10:00 a.m. on Saturday in Washington, D.C. and Gavin Reynolds—the annoying senior F.B.I. agent—had been busy. His pesky agents had been watching April's actions closely. They knew she was attempting to get search warrants on the game makers of Rage. So, Reynolds leveraged some favors with his Assistant U.S. Attorney and one of his other buddies, who also happened to be a federal judge. He secretly obtained a search warrant, assembled a team, and flew overnight to Orange County, CA, to make an early Saturday morning raid on Rage's company headquarters in Costa Mesa, CA. Ever the showman, Reynolds tipped off a local television station, so a live feed was broadcasting to Los Angeles area audiences when they arrived at 7:00 a.m.

Reynolds grabbed a handful of local agents in Orange County to assist his team. April had a long-time friend in the O.C. office, whom Reynolds did not select as part of the search. Instead, she contacted April for some background, not knowing April was an integral part of the case.

April was making her first trip back to the office since the attack on the school, so she was heading in late on Saturday to ensure her boys were comfortable at home with Reggie. She was halfway down Dulles Parkway when she saw a call coming in from her old friend.

"Myra! Wow! This is great! Are you still in Orange County?"

"Yes," her old friend said. "I'm so sorry about your boys' school. It's frightening, I'm sure."

April affirmed the sentiment. "It's terrifying to be on the parent side of this news. Nothing could be worse. But we are all good and ensuring our children get the support they need."

"I'm sure," Myra said. "Hey, do you know some guy in D.C. named Gavin Reynolds?"

"Sure do. I often wonder if he works for our side or just his side. He's always looking out for himself. Why do you ask?"

"He grabbed a bunch of our agents this morning for a search on a local company—Rage. Do you know anything about it?" Myra asked innocently.

April sped into the side of the road at top speed and fishtailed as she slammed on her brakes.

"That bastard! He's a bastard! Are you there at the Rage headquarters?"

"No!" Myra shouted back, slightly defensive, "I'm on another case. It's broadcasting live on our NBC affiliate out here. Is there something you want me to do?"

April was ready to explode and thought her friend, Sophia King, could get her the necessary information. "No, thanks, Myra. I owe you. I owe you! Thank you for thinking of me. I need to call my Assistant U.S. Attorney."

With that, she immediately hung up and started scrolling through her contact list to call Sophia on her cell phone. Before she could make the call, a hyperlink appeared in the notification bar on her phone with a text from The Dancing Cowboy.

She immediately clicked on the link, thinking Leonardo had some gem of information. Her phone opened to the live feed transmitted by the local NBC television crew. A small crowd of mostly Saturday morning joggers was in the frame, standing in front of a low-level building with the name Rage near the roof on the third floor. One person in the crowd was Leonardo, dressed in a knit hat and baggy pants. He was wearing a Van's t-shirt and holding a skateboard. He turned and faced April with a smile and a wave.

"*Ciao*!" Leonardo chirped.

"Don't *Ciao* me!" April growled, "Did you know about this?"

"I learned of it when the news report started broadcasting. I wanted to come to watch the comedy play out," Leonardo said as he inspected one of the crew's cameras on its tripod.

"It's not funny!" April screamed, "I can't stand Reynolds!"

"Well, then you will be quite pleased to know that Gavin Reynolds, this man you work with, is chasing fool's gold."

April was confused and assumed that the Great Da Vinci had miscalculated.

"What are you talking about? This company, Rage, is at the center of this investigation."

"That would be true, if it wasn't exactly what the conspirators want you to believe," Leonardo said as he looked closer at the lens of the camera he was inspecting. "The people in that building have no knowledge of terrorism, but we have not spoken in 48 hours, so how would you know this information?"

"What?" April asked blankly.

"Dr. Freud and I have been collecting names of rebels in the D.C. area by promoting a new game character I created. It's quite a remarkable character; it has powers across many game platforms."

April was impatient, bordering on angry. She had no time for Leonardo's new-found pride in video game development.

"Look, I don't think I care about your galactic powers."

"Ahh," Leonardo said, now looking at April directly, "but those 'galactic powers,' as you say, have opened a door in Eastern Europe that is most relevant to your investigation."

April was curious. "What does a game character have to do with an Eastern European break in this case?"

"Perhaps now you will listen as I explain." Leonardo began painting a fantastic picture for April. "As I said, the game character I developed, appropriately named The Dancing Cowboy, has gaming powers never before seen in this industry. On a hunch, I created a LinkedIn profile as the developer of this new Dancing Cowboy character and said I lived in Venice, Italy. Venice is a lovely town that I always wanted to visit—"

April cut him off, "Can we talk about gondolas later? Get to the point. Please!"

"*Allora*," Leonardo sighed, "a gaming programmer in Zagreb, Croatia, contacted me. He told me he was the original programmer for the game, Rage. But of course, I already knew that. He said he loves my Dancing Cowboy character and wants to hire me for an upcoming job. I told him I was not interested, but the more I pushed away, the more he told me as an enticement to join him."

Finally, April was ready to explode with Leonardo's long wind-up. "And now you know what? Can you please get to the point?"

"Okay. Well, then. I already shared the point," Leonardo pointed to the Rage headquarters, "this company here has no connection to terrorist activities. While this is a real company, it is merely a profitable marketing arm of a sinister, deeply rooted anti-American consortium in Eastern Europe."

"Are you sure about this?" April asked.

"Quite sure," Leonardo said as he inspected a roly-poly bug he had picked up.

"But how are you sure about this?" April pressed him. "What more do you know? Get to the point, Old Man."

Leonardo obliged. "As I said, the more I rejected the programmer, the more he tried to entice me with promises of wealth and power. Finally, he admitted that he was not in the top ranks of this consortium, but he wished to be. He said I would make him more valuable, so luring me was his top priority. With each negotiation turn, he added more color to this picture. The game Rage is indeed intended to normalize rioting. There is, however, no direct connection between your case and this company.

"Your associate, Gavin Reynolds, took the bait, which has undoubtedly alerted the Eastern European group that law enforcement is a step closer. This silly broadcast is likely being played in their creepy headquarters. The center of this conspiracy is in D.C., not here in Costa Mesa. Your friend, Gavin Reynolds, will chase shadows while we solve this case together. He will look quite silly in the end."

April laughed and slapped her leg.

"I thought you said this was not funny?" Leonardo asked.

"Knowing Reynolds will pay the price for his mischief and that we will solve this case makes me the happiest F.B.I. agent in the country. Thank you!"

"*Prego*," Leonardo said in return.

"What do we do next?" April asked.

Leonardo was happy to lay out the next steps. "I expect Dr. Freud will have pinpointed the rebels we must watch most closely by late tonight or early tomorrow morning. We should plan to speak then."

With that, April got back on the road and headed cheerfully into her office.

Marko and Vasko arrived in the U.S. with instructions to increase the intensity and effectiveness of Mina's planned attack. The NATO attack they supported days before failed for a few reasons. There was some leaked intelligence, but the leaks were a function of clumsy amateurs trying to act like big shots on some social media channels which were monitored by law enforcement. The attack was ultimately stopped because law enforcement was prepared and strategically intercepted a car loaded with explosives.

Those mistakes would be avoided at the Capitol. Now that the F.B.I. had searched the Rage headquarters, the timeline for the planned attack was moved ahead. The Serbs outlined the modifications to the attack with Mina, who agreed it made sense, but the loss of life would be much higher than she had planned. She only struggled with the dilemma for a short time before she embraced it. She was a mercenary, so getting paid was her only concern.

The Serbs needed a truck or a van for a late-night rendezvous, and Mina had arranged it. In the early afternoon, about 2:00 p.m. on Saturday, she bought an old, white panel van for cash from a sketchy car dealer. There were a few questions regarding the title of the van and even fewer on how the title should be held for the new owner. She just needed a van for a few days with no strings attached. A rental car agency would create a trail that would ultimately lead back to her.

Now provisioned with wheels and some time on their hands, Marko and Vasko headed out. They drove a couple of hours to Deale, MD, searching for a bar. Nothing would be better than eating and drinking while they waited.

Deale was one of the sleepy small towns on the Chesapeake Bay with boat docks. The Serbs had received instructions to meet a boat at a pier in Deale at 1:45 a.m. on Sunday to pick up lethal cargo. They would receive exact coordinates closer to the pick-up time. They had to be nearby for a quick pick-up, so camping in a bar seemed like the best option. They could not have been happier.

Mina cautioned them sternly. "If you get picked up for anything, you're on your own."

The Serbs laughed it off and drove away. In some ways, Mina hoped they'd be picked up, so she could get rid of them. She needed to be more on board with the changes they brought with them.

The Deale area had multiple water inlets with piers on the east and west sides. Hence, the Serbs would need to navigate the town quickly.

Circumventing bridges and local streets in a small town they had never visited would require extra time. Setting up shop in a strategic spot would make things easier. Once they arrived, they drove around for an hour, arguing about the best place to wait. Then they argued about which restaurant would have the best food. Finally, they settled on the Happy Harbor Restaurant, which had a patio overlooking the water.

They used the translator on their phone to order their food and drinks. Their cheerful waiter wanted to be welcoming, so he tried to engage with them using his phone translator. But unfortunately, the Serbs just ignored the kid who thought it was fun to have foreign visitors.

They finished their first beers and held their bottles high in the air, expecting immediate service. The waiter didn't see them, so Marko banged his bottle on the table to get his attention. At this point, everyone noticed him. Marko looked at the patrons who were staring at him with contempt for his rudeness. The more they drank, the more irritable they became, which made the other customers uncomfortable. The restaurant staff hoped they would finish and leave, but that didn't happen.

Finally, the manager had to come over and ask them to leave via his translator. Instinctively, they wanted to rip the place apart, but they had business in the area. They tossed money on their table and walked out. It was just enough to cover the bill, so there was no tip. The other diners were so happy they left that they chipped in and gave the waiter a 50% tip for his ordeal.

The Serbs went to a convenience store to buy beer, drove down the street, and waited in the van.

At 1:20 a.m., the Serbs got the message they were waiting for. A boat was nearby, and the captain provided them with the location of the port he planned to dock. They knew the drop-off needed to be completed quickly, in a military fashion, so there was no time for delays.

They found the pier with minutes to spare and waited by the water. A 35-foot commercial fishing vessel slowly puttered up to the dock within minutes. The only person on board was a weathered, overweight captain. Marko jumped on board and tossed a rope to Vasko, who quickly tied it down. Marko dropped several dock edge torpedo bumpers off the side while Vasko ran to the rear to tie it down. The boat was secure within seconds.

The captain cut the engine and yelled to Marko, "You boys know what's in these crates?!"

"Yes, cargo," Marko growled.

"But what kind of cargo is it?"

Marko was trying to ignore the captain as he and Vasko tore back a canvas covering stacks of crates that were the size of apple crates.

"Yes, cargo," Marko grunted as he picked up the first three crates to unload them on the dock. However, the captain wanted more information. "Do you men speak English?"

The captain was like a fly buzzing around Vasko's head, so he yelled, "Do you want to die?"

Well, they did speak some English.

"Hey!" the captain yelled, "I'm just delivering this stuff. Easy, guys."

Marko and Vasko carried four crates simultaneously to speed things up. Marko tripped on an exposed rope on the deck and almost dropped his containers. He was irritated and a little drunk.

He motioned to the far side of the parking lot and said, "If you don't want to die, stand there."

That got the captain's full attention. He quickly hurried over and off the boat.

"Smelly, European trash bastards," he muttered as he abandoned his ship.

The first 40 crates were unloaded and jammed into the van. Each had two blue stripes painted on them. Vasko, sweating profusely, returned to the boat and pulled back another tarp that covered ten more crates; however, these ten had two red stripes. The Serbs flung them around with the same disregard as the others. Finally, they put all the crates into the panel van, slammed the rear doors shut, and drove away without saying anything to the captain.

The captain tossed his cigarette and quickly shuffled back to his boat. He wanted to get away as fast as possible.

"Jesus, what was that? This job isn't worth it," he muttered.

He untied the lines, pulled in the bumpers, then looked north and south to see if any other boats were on the water. Everything was quiet, so he pulled away slowly, with only his red and green navigation lights turned on.

———◦———

The Serbs returned to the RACS warehouse at 3:45 a.m., where Mina met them. Marko opened his door, and three glass beer bottles fell on the driveway. Mina unlocked the warehouse door and directed the men to stack the crates in an unused closet.

"Were there any problems?" she asked as they opened the van's rear door.

Marko was now tired, as well as drunk. He just wanted to go to sleep.

"No!" he barked impatiently. Mina inspected the sealed crates.

"How do these things work?"

"Blue stipe cases have small glass paintballs with red paint. Red stripe cases also have small glass paintballs with red paint and sarin gas. They break, and people die. Not complicated," Vasko said as he carried some red-striped cases to the closet.

The Serbs were carefree with the crates. They didn't seem to have a death wish, so they must have known the containers were safe to toss around like rag dolls.

Dart had grown suspicious when the Serbs arrived. Mina's comment about her backer becoming more intrusive as he increased funding made him nervous. So, he had installed a few small wireless cameras around the warehouse, recording video to one of his cloud accounts.

Fortunately for him, it was only hours before he saw the trio loading 50 crates into the closet. Mina had installed a lock on the closet door to make matters more alarming. Only Mina had the combination. No one would usually notice or use that closet; it was out of the way. She picked it on purpose, hoping to keep the crates stored without anyone's knowledge.

6

HISTORY'S GREATEST SECRET

A Small Country to the Rescue

Although last week's Greek Easter celebration was an annual highlight, it cut into George and Spiro's favorite part of each week: Sunday morning breakfast at the Plaka Grill, usually at 10:30.

When George and Spiro sold their restaurant a couple of years ago, Saruha, their long-time friend and manager, wanted to buy it. Unfortunately, it was a hefty transaction, and Saruha didn't have the capital to pull it off, so there were some tough days. He needed to work and was capable, but neither George nor Spiro wanted to be the bank for such a significant transaction.

To be clear, George and Spiro would have financed the whole thing for Saruha, but their clear-thinking wives stepped in and engineered a better solution. It was a $4,000,000 transaction, and after some debt settlements, Saruha's earned equity in the sale would net him over $150,000. George and Spiro had granted Saruha ownership shares over the years but retained over 90% of the business interest.

George and Spiro's wives suggested that it was best to cash out rather than finance—and put the entire $4,000,000 at risk. Their plan called for each family to invest $150,000 and open a new restaurant. With this, everyone would be a winner.

Thus, they opened Plaka Grill, and Saruha was the owner/operator. George and Spiro only had one requirement for their investment, that Plaka Grill would not open for business on Sundays. This allowed the men to come in after church and have the restaurant to themselves for private Sunday morning feasts. George and Spiro did all the cooking, which they loved. With the community still shaken from the near disaster at the elementary school, the men sought refuge at their favorite spot.

It was a small club they had started, and every Greek man within 50 miles would have gladly come to enjoy a small fraternity like this. However, it was only the backgammon group today, so Costa Popoulos also joined. Saruha had configured his television monitor and a camera to include Da Vinci and Freud. They appeared at a cafe in the Plaka district in Athens, Greece. Leonardo had abandoned his ancient Greek fisherman's outfit. He chose something more contemporary to look like he was coming from a local church. Freud, well, he always wore the same thing, but it was his look.

The men always celebrated Costa when he was in attendance. They treated him with great respect and admiration. He was exceptionally well educated and worked in the Greek Embassy, so he held a special status with Saruha, George, and Spiro.

While Costa sat enjoying his first cup of Greek coffee, the others were busy cooking in the kitchen. Leonardo asked a question. "Costa, my learned friend, what does the Greek word *Kryptos* mean to you?"

Freud leaned in with a keen interest.

"Well," Costa replied, "it's an ancient term that we hear less often, which means 'hidden secret,' and was somewhat popularized in..." Costa stopped, then smiled, and continued, "Popularized in *The Da Vinci Code*, using a device they called a Cryptex, which derives from the Greek word *Kryptos*."

Leonardo folded his legs and looked at Freud with great satisfaction, then added, "That is correct, and there is an art structure named *Kryptos* in front of the CIA!"

"Yes, the house of secrets, as we call it. These interesting facts make me wonder about our current circumstances," Costa said as he scratched his chin. Da Vinci and Freud said nothing. "Curiously, April Falk works for the F.B.I., an art structure at the CIA is named *Kryptos*, and you are discussing a famous movie that included a Cryptex."

"The *Kryptos* at the CIA takes some artistic license with the word *kryptos*. There are secrets in the art structure, but they are not hidden at all; they are there in plain view. A hidden secret is one that the world never knew existed," Leonardo said as he pulled out his *komboloi* - Greek worry beads.

"And what of the *Da Vinci Code* author, Dan Brown? Did he know some hidden secret he wanted to share?" Costa asked curiously.

"I will investigate that," Freud interjected, "but, for the moment, I am comfortable that he wrote a provocative story woven with history."

However, the great psychologist was anxious to get to his point and had more questions for Costa.

"We have been exploring this notion of a *Kryptos* - a true hidden secret that connects us in this unique relationship. It is a bit of a mystery to us, but I have a hypothesis."

Costa and his Greek friends had also been sufficiently curious about the familiarity of history's geniuses. A discussion on this topic fascinated him. Leonardo smiled and slightly waved his hand to bid Freud to continue.

Freud began to probe. "If I may ask, please, your senses prompt memorable images of your life, don't they?"

Costa smiled widely. "Why yes, just the other day, a refurbished 1971 Mustang pulled in next to me at the gas station. I immediately thought of my younger days as a struggling Ph.D. student. But those were very happy days for me, and reflecting on them gave me great joy. I told the owner his car was an extraordinary time machine."

Freud advanced the topic.

"You are well aware; your senses trigger your memory. So how do your other senses transport you back in time?"

Costa reflected for a moment. "Well, if you mean my sense of smell, I often encounter foods that make me feel like a young boy again. I think we all have wonderful memories of our favorite things from the past."

Freud sat quietly while Da Vinci fiddled peacefully with his *komboloi*.

"Music," Costa exclaimed. "Music transports me back. I so enjoy hearing the music from my youthful days, singing along as though I'm 12 years old once again, and racing through fields with my friends near the old Palestera in the ancient Olympic village in Olympia. The memories are so vivid; it often feels like yesterday."

Next, Freud moved to tie it all together. "Your senses directly connect you to your history and transport you through time. You seem to recall them with great fondness. Is that a fair statement?"

"But, of course," Costa paused, then continued, "So, tell me, Dr. Freud, your line of questioning is not without purpose. What are you asking me?"

"Well, it is well-known that the human brain has a vastly untapped capacity, and I believe there are some hidden secrets in the brain that can explain our relationship."

Costa was masterful at gathering information and allowing facts to be presented before speaking. So, it was not unusual that he did not pounce on Freud's profound statement. It was one of the skills that made him invaluable in the diplomatic community.

The three sat quietly for a moment, then Freud looped around to a seminal scene from the famous movie *The Da Vinci Code*. The actor, Tom Hanks, conjured mystical images of art, science, and astronomy through the ages in search of clues to a password in his attempt to open the Cryptex device and gain access to its hidden secret.

"Do you recall that scene?" Freud asked.

"But of course! That movie was well-written, directed, and acted. So please tell me, how is my brain connected to the *kryptos* depicted in a popular film?"

Freud answered him, "Think of it as a metaphor. The imagery that actor Tom Hanks envisioned led him on a short journey of enlightenment. Those images prompted his mind to reveal an important password correctly. It was theater, but it was a great depiction of the question I am about to ask you."

Freud began to speak in a pleasant, mesmerizing, and relaxing tone. He spoke of some of the hidden secrets throughout history: Stonehenge, alien life, the pyramids, and many more. He revisited the hypothesis presented in *The Da Vinci Code* movie—the hidden secret that a living descendent of Jesus Christ was among us.

Freud led Costa with a line of questioning to stimulate his mind. Neither Da Vinci nor Freud had been able to solve this puzzle. It was the apparent first flaw in their AI coding—a human brain was required to make these extraordinary connections.

Freud had previously told Da Vinci that he believed there were individual dormant memory neurons unique to Greeks' brains, which had been somewhat aroused when they met. Moreover, he speculated that these memory neurons were more prominent in some Greeks and less so in others. So, there was no rhyme or reason that a random sample of Greeks would be familiar or unfamiliar with Da Vinci and himself. Freud was convinced that stimulating certain neural transmitters using this meditation of history might create enlightenment with someone as brilliant as Costa.

Freud had shared with Leonardo that he believed it would be critical to quantify this mystery. First, it would unlock the most significant psychological development in history, and second, he was concerned about other current-day societies. So, he was thinking ahead, but knowing how to manage these memory neurons would be helpful in the future.

April, who was not Greek, was different. She didn't know either Da Vinci or Freud, but reluctantly, she began to build a normal relationship.

The Greeks were a different story.

Freud prompted Costa to explore the depths of his extensive education. He spoke of the teachings of Socrates and Plato, the mysteries of the gods, the foundation of government, and Greek engineering marvels. Freud meticulously built a mental puzzle connecting all of Costa's life experiences and life's learning.

Then Costa responded and spoke of his childhood. Next, he talked about the history of Greece and how it was passed on generationally for safekeeping, as it has been for ages. Costa had always known that Greeks had a special connection to history, somewhat unique to other cultures. His time working in the diplomatic corps had confirmed that.

Costa envisioned lineage images: the Popes, the monarchy, the history of American presidents, Greece's storied history, and even his family tree. Each represented a generation full of knowledge, history, and advancement. Generations shared a lineage, but not a personal relationship. Once people leave this earth, their loved ones and scholars are in the business of remembering.

Then, in an instant, something clicked inside of him.

Costa felt as though his body lit up with a great inner light. Da Vinci and Freud looked upon him intensely as his face began to shine. Was there another great secret in this world, never known or discussed? Did the world's greatest invention unveil a mystery of the ages? It was as though all the doors began to unlock. Costa felt a strange confidence that he was finally starting to understand that there was a mystical connection to history.

Costa looked into space as he saw a million puzzle pieces of past and present lives begin to connect.

"Greeks have an unbroken relationship with history. That is why you were not strangers to me when we first met. I had never met you, but it felt like we had always been friends. It was as if I had not seen you for an hour or so. I see it now! I don't have an unbroken relationship with you, but you are a conduit to history, and history pulses through our veins like an instinct."

Freud folded his arms, raised his chin proudly, and tapped his foot on the ground. "Precisely. It is instinctive for Greeks to connect with history. This is exactly what I suspected. When we have time, we must explore the genetic code of the Greek society."

In astonishment, Leonardo shook his head slightly in disbelief, then looked at Freud and said, "You were correct about this wise man."

"Indeed, I was, and more precisely, I was correct about this wise Greek man."

Almost immediately, Leonardo tried to move the conversation forward. With a magnificent box checked, he was ready to move ahead. Uncharacteristically, Costa impulsively stopped him. "But it's not every Greek. So why is it some Greeks and not others? What is the hidden genetic code you speak of?"

As was always the case, Leonardo wanted to move very fast, and today's historic revelation was merely a steppingstone. Talking about every Greek was not the next step in Leonardo's plan. He had aimed to expose and answer a mystery. The notion that everyone should stop and stand in awe must have been another missing piece of his code. Or was it?

"We can explore that another day, my wise friend." Leonardo pressed forward. "You have a great responsibility ahead of you. This technology—which we have called 'Pali'—must be protected from those who want to change the future of the civilized world to gain power and control. So many evil forces are coming, and the Greeks are the great Centurions. I have been quite sure of this for some time. You are the keepers of history; that is the true revelation of this *kryptos*. The civilized world relies upon you to provide a safe harbor as we develop this technology for the good of all."

Costa sat up at attention. The great Leonardo Da Vinci just called upon Costa and his fellow Greeks to take a stand. To help shape the world's future again, as the Greeks had done over 2,000 years ago.

"Is this a remarkable new renaissance you commission upon the Greeks?" Costa inquired.

Leonardo put his *komboloi* down, then stood up. "The advent of this incredible technology will prompt significant change. The outcome has too many variables that we cannot control. So I will not speculate on the future. The work ahead involves laying the necessary groundwork to fend off bad actors, while influencing responsible behaviors for the civilized world. We dare not fail."

Still sitting at attention, as though he was addressing a group of world leaders, Costa inquired, "What would you have me do? How can I help you?"

Leonardo sat down and picked up his *komboloi*.

"Think of our journey as a critical path - or perhaps more commonly, like a string of Christmas tree lights. In the past, if one light anywhere in the circuit went out, the entire strand would go dark," Leonardo said as he explained the journey forward. "Every action ahead of us could take down our mission to develop this technology, which would lead to more significant problems in the world. So, April Falk must be the center of our world right now. She must remain as the tip of our spear. To do so, we must help her solve the conspiracy against her government. Once we unravel the criminal conspiracy, we will then need the world of diplomacy. Your diplomatic role in this mission will be as vital as April's role in solving this case."

Based on Leonardo's vague explanation, Costa needed clarification to understand the big picture.

"I work in the embassy of a small country that does not wield great power over this great nation. How do you propose Greece help in this endeavor?"

"I have three plans in mind, but I don't know which of the three we will pursue yet. There are a variety of outcomes still possible in April's case," Leonardo answered confidently.

He saw George coming out of the kitchen.

"For now, eat. Then after we eat, we dance. Another day, we will talk about this *kryptos*."

This was all within the earshot of George, who heard the word *kryptos*. Costa was savvy and patient enough to let the dialog meander in another direction.

"*Kryptos*!" George exclaimed. "Did you two watch me make my secret pancake batter?!"

George had used the term kryptos to fend off customers asking for his secret pancake batter for 40 years.

"No, George, the secret of the batter is safe," Costa said. "We were talking about the *Kryptos* art at the CIA."

Every Greek symbol or word in America made George proud.

"I have heard of that *Kryptos*, but what is this talk of dancing after we eat? Which Greek dance shall we dance? Shall I get the music for *Sirtaki*?" George asked.

Leonardo stood and raised his arms. "Today, we dance the Electric Slide, and Sigmund Freud will teach us."

Freud displayed an uncharacteristic smile. "I have been practicing. We will have great fun with this happy dance."

Spiro and Saruha came from the kitchen with trays of meats, cheeses, and olives. Then they all sat and had a feast.

7

COUNTDOWN TO ATTACK

Lost and All Alone

Da Vinci and Freud's Saturday night was fruitful.

It was 11:00 a.m. on Sunday. Leonardo created a campaign targeting D.C. gamers, which worked to perfection. He pinpointed local gamers based on the data previously collected, then promised the target group more free Dancing Cowboy powers. They had every reason to believe they were randomly selected but wouldn't have cared otherwise. The excitement around this new gaming tool was frenetic.

An emoji of the Dancing Cowboy scrolled into their text messages on their phones. Once opened, a quote bubble above read, "Click me if you want more Dancing Cowboy powers!"

That was too easy. Everyone Leonardo targeted clicked on the link. Then, the instructions began to pop up, outlining how each gamer could claim additional free powers. In return for the new, free capabilities, each recipient needed to answer a survey as part of a "research" project for the game developer. Everyone eagerly accepted.

Initially, when the survey opened, it appeared to be an interactive video, but it wasn't. Instead, it was a live feed with Freud, who acted a bit robotic for effect. Freud, dressed in his formal attire and seated at his desk with many books on shelves behind him, opened each chat session with a greeting.

"Hello, this survey will only take three minutes. After completing this session, you will receive all the new Dancing Cowboy powers."

The gamers gladly participated.

Freud had already learned that gamers have common personality traits that were easy to exploit, enabling him to gain more insights about each person's protest group. They spent hours upon years playing video games. Knowing so much about their psychology allowed Freud to quickly put each gamer into a meditative state of relaxation. This slumber enabled him to gather needed intelligence about protest groups and the method and manner with which they communicated.

Freud learned a critical aspect of how gamers planned protests after ten or so interviews. An essential element of monitoring gamers' chatter was an invitation into a seldom-visited sub-chat room. Freud needed to get himself or Da Vinci into that chat room. Hearing this, Leonardo quickly added a "waiting room" to the survey experience, so gamers chatted with one another while waiting to be called in for the survey. He put himself in the waiting room, then chatted it up with gamers, befriending them.

During his interviews, Freud allowed gamers to take a few minutes to recall pleasant memories long since forgotten. The gaming culture creates not-so-subtle anxiety and blocks each person from dreaming, remembering, and exploring. Freud's soft, easy tone naturally invited the subject to slow down and think about their days before gaming.

He told Leonardo that it was like being let out of prison after many years to run in a grassy field on a beautiful spring day. Each gamer had forgotten the pleasures of life without electronics because of their anxiety to play, achieve and win. This state of elevated consciousness and freedom allowed Freud to ask probing questions.

The gamers experienced blissful relaxation, lowering their guard about passwords, planned raids, and leadership structures. Freud conducted over 700 live video chat sessions with D.C. area gamers. As Leonardo had previously told April, it's not a violation of the law if people are willing to give you the information you seek.

The virtual dream team now had a good picture of the Washington, D.C. protester sub-culture. So, they built a report for April, outlining the names of the 14 most prominent groups, where they hung out, who their leaders were, and how they typically communicated ahead of protests. Despite Leonardo's earlier focus on a need to access iPhones and devices for intelligence, the team learned that they only needed to build profiles on a few seldom-played game sites. These lonely game sites had chat rooms that law enforcement never looked at because the kids who typically played them were ten and under. Leonardo had also cajoled his way into the comfort zone of some gamers who'd promised to invite him into the D.C. area sub-chat.

It was simple. Now they had access and could monitor chat among the D.C. protest groups.

The RACS protest group was on their target list but didn't stand out meaningfully. So, the database report they developed for April included Dart's name, but it was no more significant than 40 other names on the list. Mina's first name was on the list but remained off Leonardo's primary radar. She had no online presence, so there were no bells and whistles for anyone to look at. No last name, no footprint, just a ghost. Leonardo noted that her lack of any digital fingerprint was relevant, and he tucked it away. The overlapping word "Rage" in the game and the local protest group didn't even raise a flag with Leonardo, but that would change.

At exactly 1:30 p.m. on Sunday, Mina met Marko and Vasko back at the warehouse. They needed to transport one of the dangerous blue crates to a rural farmhouse west of Washington, D.C. Mina liked the curve balls the Serbs were tossing into her planned attack—her revised plan called for adding a case of lethal paintballs to the bomb she was building. The paintballs were the size of large marbles, so one crate of them could take down thousands of people, depending on the concentration of people standing together. A bomb with deadly gasses would achieve destruction, death, and a large quarantine radius. As the U.S. Capitol was her intended target, Fonok personally engineered the modification to keep the businesses in the Capitol closed for some extended time after the attack. It would be the ultimate strike of an anti-capitalist upon America.

They loaded the crate into the white panel van, and Mina set out driving on Route 66, the major highway east and west in the Metro area. The small, unincorporated area in Albin, VA, was vast farmland and trees situated 80 miles west of Washington, D.C.

Unfortunately, the Serbs were always hungry, so she had to pull into a McDonald's for three bags of burgers and fries. Mina scolded Marko for tossing his trash out of the car while they were driving. She didn't care about the environment, but she didn't want to draw attention in any form.

After another 25 minutes, Mina made a final few turns on roads with no signs, then turned onto a private gravel road. Nearly a quarter mile up, behind a thick cluster of trees, was a run-down farmhouse next to an old barn that looked like a junk yard for auto parts.

There were torn-apart cars everywhere. Someone had set up a large, portable canopy in front of the barn covering a gleaming new white Chevy Suburban. The trio pulled the panel van up past the Suburban and got out. Mina ran off quickly to the barn while Marko and Vasko stretched and looked around. They could tell it wasn't a working farm because there were no animals, fences, or farm equipment.

Less than a minute later, a big voice boomed out in Serbian from behind the Suburban.

"Brothers!" an old man yelled, "Welcome to America!"

The Serbs looked at one another in surprise, knowing a fellow compatriot was there. The older man was Zoran. He popped out from behind the Suburban with the help of two canes. His left leg dragged, but it didn't slow him down. He was quite mobile as he sped up to the two. As he got closer, Vasko noticed a scar on the left side of Zoran's face, and his eye was a lousy prosthetic. Then Marko recognized him.

"Zoran! Zoran, it's been a long time," he exclaimed in Serbian.

Mina was still in the barn as the three men wandered over to a group of old chairs under a nearby gnarled tree for a reunion.

"I know everyone was wondering where I went," Zoran said, the conversation taking place entirely in Serbian. "The U.S. State Department let a big refugee wave in through Kosovo. They flew us all to New York. I got papers, came here, and bought this farm with the money I saved over the years."

Vasko was indignant. "The money you saved?! Old man, you make me laugh."

Zoran was a well-known, well-paid bomb maker during the Kosovo War. Unfortunately for Zoran, a NATO airstrike obliterated his workshop and severely injured him. It was his good fortune that peacekeeping forces were nearby and evacuated him before they realized that his old house was a bomb-making nest. He received treatment at a military field hospital. After recovering from his injuries, the peacekeepers moved him in with a refugee group bound for the U.S.

"I dared not tell anyone where I was. I had a good cover and a ticket out. I didn't want to ruin that and get sent back for the war trials," Zoran added.

Marko and Vasko had a more challenging time as that war ended. They were part of the corrupt Serbian Special Police force that had to scatter when NATO peacekeeping forces arrived. Their group was trying to travel north to southern Hungary, hoping to reunite with others who preceded them in escaping. Instead, their numbers dwindled each day as peacekeeping forces hunted down the fleeing troops.

Nevertheless, they made it through and eventually connected with a mercenary team. They had been in what they called the 'dirty business' for the past 25 years. The contractors in the dirty business were former military and terrorist personnel. If you were a drug lord, anti-capitalist, terrorist cell, or even a rogue team in a legitimate government, you hired these thugs for jobs.

Mina had started out in this loose confederation but was now on her fifth job with Fonok, who kept her busy and well-paid. Fonok previously paired her with a small team that botched everything. She was on the run for a month, trying to distance herself from that experience.

Zoran kept up with the news about where the escaping military forces went, what they were doing, and if they, like him, were hired for notorious jobs. He knew big jobs that paid well, like Mina's, were hard to come by. The dirty business contractors often took nickel-and-dime jobs by comparison. It was all they could get until they made their bones on big jobs.

"They will pay you well for this job. Then you can come to live here with me," Zoran said as he proudly pointed to his farm.

Mina exited the barn and walked under the canopy to inspect the Suburban.

Zoran yelled to her from a distance, "These boys will live here with me for a while after the job."

She laughed at the notion that three fugitive Serbs would live in plain sight, only a short distance from the U.S. Capitol. "You don't think you're going to stand out?" she shouted back rhetorically, "Seriously, don't you know that every law enforcement agency will be searching for you?"

Zoran dismissed her. "There's nothing here. No one comes out here. We could live here for years. Plus, they'll be looking for you anyway."

Zoran had a point.

Mina would be driving this bomb on wheels up to the U.S. Capitol building, but she had a well-thought-out escape plan.

"I'll be out of the country before they put out all the fires."

Vasko was curious. "How are you getting away so fast, and where are you going?"

She laughed off the question. She would have to keep her plans private from this group for fear that law enforcement would capture any of the trio. Mina was a lone wolf—with a plus one.

Zoran got up, aided by his canes. "Nothing is happening, and no one is getting paid if we don't put all those devil balls into this van and get this thing ready. So, let's finish up."

They wandered over to the Suburban. Vasko was curious about the bomb's impact.

"What is the lethal range of this bomb?" he asked as he looked inside the rear door.

"About 1,200 yards," Zoran answered proudly. "If she gets it in the correct position, it will clip off a big section of the Capitol and maybe the Supreme Court."

Then, Zoran asked, "Are these devil balls going to work?"

Marko was nodding his head and laughing a bit. "This residual gas will shut down the Capitol for weeks. And if all those extra blue balls get tossed around on the west side of the Capitol, bodies will be everywhere. So, our strike will be big, Fonok will be happy, and we will all get paid."

Zoran had worked within this consortium previously and was always a little suspicious of Fonok and Mina.

"How much have you been paid already?" he asked the two Serbs.

"About 25%," Mina answered quickly, hoping to stop the trajectory of this discussion. Zoran knew the Serbs were at risk of not being paid.

"The boys should collect another 60% before you deliver this van. You know these boys will have a hard time getting paid after an attack like this," he said, looking at Mina with disdain.

She wasn't happy that the old man was making her life more difficult, but there was no honor among thieves. Marko and Vasko looked at her like two dogs eying a piece of red meat. The two knew they had some control over the situation because they needed to secure the Suburban for a few days in the D.C. area. Mina was anxious to finish and knew she was in a corner.

"Okay, okay. I'll get this worked out when we get back. Let's get to work here."

The four of them labored most of the night under the bright lights in the canopy. Zoran had removed all the rear seats and floorboards to pack the bomb into the extra-large cargo section. Then, finally, the

Serbs carefully loaded the paintballs into the Suburban. Zoran and Mina backed away when they began. Marko and Vasko didn't have the same concern. They had worked with these paintballs many times before and knew that the glass coating was sturdy enough not to break unless thrown with a bit of force.

Once they finished, Mina tinted the windows so that no one could see the massive bomb inside. The men were exhausted and sitting in some chairs when Mina returned to the panel van and pulled out some sizeable magnetic car decals. She said nothing and held one of them up, showing them off to the trio as she walked by.

"As soon as you guys told me about the paintball plan, I figured this would give me greater access, so I made them up yesterday."

All the magnetic decals were exact replicas of the Washington, D.C. Police Hazmat Squad.

The white Suburban was the same model used by the Brass in D.C. Now it became apparent why Mina liked the Serbs' curveball; she was counting on chaos ensuing if sarin gas paintballs got tossed around at the Capitol. An emergency of that magnitude would flood the network with distress calls, and one more D.C. Hazmat truck would have easier access to the area of the Capitol building. There was unanimous agreement. This plan would be highly effective.

The Suburban was ready with the decals tucked inside the back door. Zoran walked through the detonation sequence using a dedicated cell phone he had prepared. "You need to be almost 1,200 meters away before you detonate. How fast can you cover that distance on foot?"

Mina was fit, so getting away fast wasn't her big concern. However, she knew she had to allocate some extra time to navigate the pandemonium and escape the conflict zone. Otherwise, she might stumble into a problem. "If we plan this right, we can have all the attention on one side of the Capitol, so a Hazmat truck on the other side won't raise any alarm. I have at least seven minutes before I need to detonate this beast. After that, I'll be fine."

It was just before sunrise on Monday, and the four mercenaries had fallen asleep in their chairs for a few hours with some old blankets wrapped around them. Mina was the first to wake up, and she roused the Serbs to get moving. Their empty beer bottles on the ground banged together loudly as they quickly shuffled to their vehicles.

They needed to get the Suburban back to D.C., pick up the other 49 crates of paintballs, and move everything to another storage location that she had rented. She couldn't park a shiny new Suburban near the RACS warehouse. Mina drove the Suburban, and the Serbs followed behind just ahead of the impending Monday morning traffic jam on Route 66 into the District. They made a beeline back to the RACS warehouse, arriving at 6:40 a.m. Much to her chagrin, all the neighboring contractors were there. It was a new week, and contractors always started early.

Mina didn't know much about contractors' lives. As much as she didn't want to stand out, she had no choice. She wanted to make only one trip to the new storage location, so they pulled in as the next-door neighbor, Eduardo Gomez, loaded water coolers onto his trucks. It was hard not to notice a woman and two brutes pulling into the lot in a shiny new Suburban, then backing a panel van into a "vacant" warehouse any time of day.

Mina said nothing and didn't make eye contact with anyone. She walked with her head down and pretended she didn't notice Eduardo watching her every move from just a few feet away. The threesome quickly packed the crates in the van, pulled it out, then closed and locked the 16' roll-up door. A local news crew had pulled in to interview Eduardo while the trio was loading the van. He was still a hot interview after last week's near disaster at the school. The sight of the crew was alarming, and Mina swore and violently pounded the dashboard as they drove away.

To make matters worse, Mina did not know that Dart had installed security cameras inside the warehouse. While Mina was out in Virginia working on the Suburban yesterday, he had seen the video of the threesome loading red and blue stripe crates into the locked closet. Dart knew she was operating independently on something, so he changed the camera settings to notify him via text if any motion was detected. So, in addition to Eduardo Gomez and a local television news crew seeing them early Monday morning, Dart rose quickly at his nearby house and raced over to the warehouse to watch from a distance. Mina's two vehicles—the panel van and the Suburban—were previously unknown to Dart. His anxiety was making him sick, but his curiosity was piqued. Some trouble was brewing. It couldn't have been more obvious.

Mina didn't know Dart was following them as they drove the Suburban and panel van to a new warehouse one mile away in nearby Bailey's Crossroads. Dart was 300 yards behind the panel van and saw it was coming to a stop to make a left turn into a parking lot at a storage warehouse. Once he confirmed they were turning in, he crossed two lanes to observe from a strip mall across the street. The warehouse complex was behind a painted black iron fence, so it was relatively easy to see them with the binoculars he'd brought. The trio wandered down the interior driveway to the third of six buildings. Dart was able to stay parallel with them from across the street. He had a decent vantage point and could see the threesome stop in front of one of the many garage doors.

Mina unlocked the door and backed the Suburban in first. Then, Marko pulled the panel van up, got out, and popped a surprise on her.

"Let's settle up now," he said sternly in Serbian.

Vasko stood nearby with his arms folded, nodding to confirm that he wanted payment also.

"What?! You think I can call and get money transferred like a bank?!" Mina yelled.

She was furious with Zoran for putting these thoughts in their heads. "Let me arrange it early this afternoon. I can set up a secure text communication and take care of it," she pleaded.

The Serbs were intractable.

"How on earth do you think I'm going to get this money transferred from here in this lot?"

Vasko opened his phone, scrolled to a screen with wiring instructions to a bank in Malta, and shoved it in Mina's face. "You can put all the money owed to us in this account." Mina was incredulous. "That's like $80,000 US!"

Marko calculated the figures on his phone. "No, it's $109,000 US."

The Serbs had built their plan on the drive back into the city from Virginia. Dart could see they were arguing from across the street but couldn't hear them over all the road noise.

Mina had a mean, evil side, but that didn't faze either man. She had no choice, as her negotiated fee was over $650,000. She wasn't going to let that slip away. She would have to front the money herself and collect the total later.

"You two are a pain in my ass. I'll arrange to transfer $75,000 now, but I need you accountable to finish this job. So, you'll get the balance as agreed after we finish."

She pretended to negotiate but would have paid them the whole amount to keep things going. To her surprise, they agreed. So, she took out her phone, and within four minutes, the money was showing in their accounts. The Serbs were happy, Mina was not, but they finished loading the crates from the warehouse and drove off together in the panel van. Dart now knew that cancer had grown in his group but didn't know how to deal with it. He had no one he could trust, so he wasn't sure what to do.

———◆———

It was 11:45 a.m. The past seven days had been exhausting for both April and Nick. Nick had to balance the emotional stress of the attack on his kids' school and the unimaginable highs and lows he faced in his relationship with Leonardo.

Of course, April's children were in the same circumstances, and her journey with Da Vinci and Freud was no less of a roller coaster ride. Moments of clarity came in waves like involuntary spasms. Then she'd have near panic attacks about shielding all this from the Bureau and wonder if she was in over her head.

She balanced the wife, mother, and agent roles well in her career. However, these new roles, imposed upon her by the galactic dream team, were unnerving. She didn't want to be an "explorer," and the notion of violating her oath of office felt like a hole in her stomach when she put all this in perspective. As bad as that felt at any moment, she'd have the security of the country and thousands of civilians' lives in her hands in a matter of days. She would face the hardest decision any President, military leader, or F.B.I. agent would ever make, and it would be all intertwined with the one and only Leonardo Da Vinci.

It was 2:15 p.m. Mina had slept a few hours but woke up steaming from the argument this morning. She wandered over to Jo's Cafe and logged on to Threema. The pre-payment issue with the Serbs was not something she would discuss with Fonok. That would only create a concern, and she didn't want to deal with something else.

So instead, she texted her "plus one" partner, Ivan. Mina and Ivan had been partners for years on jobs. Like her, he was a former militia soldier, but he wasn't a big hulk like the Serbs. Ivan was lean and not heavy, but he was all steel. He had incredible strength. Like Mina, he also had paramilitary training and was a self-defense and weapons expert. Ivan didn't stand out, and that was the plan.

The pair split up on purpose when they arrived in D.C. Their mission was to infiltrate protest groups and use them as a launch pad for an attack against the U.S. The nature of the attack had only taken shape in the last couple of months. Mina estimated there would be multiple benefits working apart, and there were.

For example, when Dart was trying to determine if he should let her take a more significant role, Mina gave him Ivan's number, who was already a big player in another protest group known to Dart. When Ivan confirmed Mina as a solid addition, Dart took his word for it. However, the biggest reason for splitting up was to create a fail-safe option for their mission. If either became compromised, the other could execute the mission. Ivan purposely maintained a mid-level role in the other group.

It kept him off the radar, but he had access to all the chat rooms and got intelligence from the ground level within the community.

Initially, the plan Mina had developed called for a massive protest on the west side of the Capitol lawn while she drove a bomb-filled car to the east side. Mina estimated the protesters would be an enormous diversion of resources and attention, creating possible security lapses. Getting a bomb as close as possible was sufficient.

There would be some destruction, but the symbolism would be massive. Adding the lethal paintballs, the Serbs delivered made the entire mission more powerful. This modification to the plan had Fonak's fingerprints all over it, but he never mentioned his involvement. He said the Serbs were coming and that getting them integrated was important. The protesters would be unsuspecting mules carrying lethal and non-lethal marbles in their pockets up to the steps of the U.S. Capitol.

Assuming they started breaking the deadly balls on or near the Capitol, there would be an emergency of epic proportions. The first calls by Capitol police would include Hazmat teams. With that level of disaster and chaos, Mina figured she'd be able to get much closer to the Capitol dressed in an official D.C. police uniform and driving an official Hazmat Suburban. That's why she liked the Serbs' plan. At least until they made her cough up $75,000 of her own money that morning.

Text sessions between Mina and Ivan were brief. They never talked on the phone. She and Ivan had already traded notes on the Serbs' involvement, so he knew about the paintballs. Now it was time to get moving. Mina opened the app on her phone and waited for Ivan to reply.

Mina: The F.B.I. raided the Rage H.Q. on Saturday. We have to move ahead.

Ivan: Okay. How do you want me to play?

Mina: Let's do a dry run with non-lethal paintballs tonight.

Ivan: At the Capitol?

Mina: No. Tell everyone to meet at 5th Street, N.W., at 22:00.

Ivan: Cross Street?

Mina: P Street, in the alley

Ivan: Want me to send the alert?

Mina: Yes and tell them I'm bringing paintballs for the protest.

Ivan: Okay. I'll send it to the chat room at 17:15.

Mina: Perfect. See you then.

It was 5:05 p.m., and April had spent all day reviewing the intelligence her team had accumulated. It was fascinating that while in the privacy of her office, she could compare and contrast the research completed by her Bureau team, Leonardo's report, and the early results of Gavin Reynolds' raid on the Rage headquarters. But, of course, she had a good laugh because Reynolds had produced nothing. Rage was a well-run company that operated entirely within the laws and was well-known in Orange County for creating a foundation to fund STEM programs for underprivileged children. It was stunning to see how different the intelligence was when comparing her team's efforts to that of Leonardo's. He and Freud monitored chat exchanges in real-time and pinpointed surveillance targets.

She wanted to meet with her team leads before everyone left for the day, so she called them into her office. She loved her team—at least the part who worked to support her, not Reynolds. They were brilliant, intense, hungry, and did everything by the book.

The first thing she did was to outline a surveillance schedule with targets, including the RACS headquarters, among many. Unfortunately, no one could find a residence for Darren Temple. Otherwise, they would have also watched that location. Their public records search failed to uncover that Temple's house was purchased in a family L.L.C. under a nondescript name, so no one knew where he lived—including anyone in the RACS group.

Temple's place was quite lovely in a nearby residential neighborhood. The image of an old English Tudor home and manicured lawn seemed entirely at odds with one of Washington's leading professional protesters. Thanks to his family, Temple was doing just fine.

April's team needed help reconciling the intelligence she used to pinpoint these surveillance targets, which perplexed them. They didn't want to challenge her, but they wondered where this newly unearthed information came from. Only some of the surveillance targets came from their intelligence.

Once April made the assignments and they were wrapping up, it was clear that she needed to provide more context around the intelligence.

"Okay, okay," April started, "you guys need to work on your poker faces a bit. I can read you all like a book. What's wrong?"

Ashlie Jackson spoke up. "Where did..." she stopped, then started again. "I can't see some of these targets on our intelligence."

She was too nervous to look up because she'd called out the boss.

"Ashlie, I'm right here. Look at me, please," April said supportively.

Ashlie locked eyes with April.

"Don't be afraid or apologetic when you smell a fish. That is an important skill. You're wondering, 'How did she develop these targets?' You're right. So go with it."

Ashlie was emboldened. She sat up and spoke directly. "Okay, so, where are you getting your intelligence? None of us have seen some of these specific targets in our research."

"Great job," April complimented, but she knew she needed to add something insightful and honest. "If it was a source, don't you think I would have received an approval from and registered it with our Asset Coordinator?"

The team nodded to affirm the statement. In 2004, the F.B.I. started a reengineering project around Confidential Human Sources. By 2007, the Bureau created the role of Asset Coordinator and conducted extensive training in 56 field divisions to ensure that a coordinated, cross-agency policy was in place to manage sources.

However, every agent knew the role of this function was a brick wall with one door. If you didn't walk through that door and follow the rules perfectly, you were considered rogue and headed for some bad days. You had to register sources. So, the head-nodding by the team was an affirmation of the culture and training within the Bureau.

Still, April needed some context for her team, so she relied on some good old "word parsing" because Leonardo wasn't human, so by the letter of the rule, he wasn't a Confidential Human Source.

"I know this technology guy who may have found a chat room where the protesters plan activities, and it's not in the Rage chat room, so that's why Reynolds' team hasn't turned anything up. This surveillance we are doing will confirm or dispel the value of this intelligence. We'll pair him up with one of our tech teams if it's valid. If not, we close the door and move on. It's worth a few hours to run this out."

That was enough. It was accurate and concise, so her team went with it and set out to get agents assigned to the various locations.

No sooner did her team leave her office than she received a text from her non-human source monitoring the chat room. Ivan's note had hit the wires, containing some curious references that Leonardo immediately saw. To be sure, Ivan had put a message out for tonight's protest, but he added that "Mina," from the "RACS" protest group, would be providing red paintballs for everyone to toss around. There was that name, "Mina."

Leonardo's ghost just made an appearance. April confirmed receipt and noted Leonardo's curiosity about this player. She wanted to get a closer look at all these characters in action, but if she wanted to be there, she would have to go alone. There would be no plausible explanation for April to know about this protest without disclosing more about her source.

If F.B.I. agents showed up at a civilian rally, there would undoubtedly be some coordination with law enforcement. That would set off bells in Reynolds' office, and she didn't want that. So, if she was going, she had to go undercover and off the radar. April also knew she would have to park some distance away and go in on foot; her car couldn't be anywhere near the protest. She didn't even want to dare take an Uber or taxi, so she knew she would have some exercise that night.

She called Reggie and told him she was tracking down a lead at 10:00 p.m. in the District, so she expected to be home by midnight. That was nothing out of the ordinary, but the wellbeing of her boys made this a different kind of night. She got them both on the phone and tried her best to be normal with questions like, "Did you take a bath? Is your homework done? Make sure you brush your teeth; there can be no gaming tonight after 8:00 p.m."

Sadly, it sounded pretty normal to them, as they were hoping for a relaxation of some rules to assuage their parents' concerns about their emotional wellbeing. A parent in a similar situation would give their kids anything after a tragedy like the one at the school. April was no different. She would have bought them new computers on the spot if they had asked, but they didn't, so she sounded just like "mom," and that's what they needed the most. It was a beautiful, calming moment, and then Gavin Reynolds stuck his head in her office. Nothing could have been worse. She waved him off with a mean look, told the boys she loved them, then called for Reynolds, who was still within earshot, so he knew her call was over.

He looked and smelled like he had just taken a shower, which he already had. He had finished his workout and made a beeline to the office to catch her before she left.

"How are your boys?" he asked.

April knew that there was only about three percent sincerity in the question, but blasting him would make her look like an ass, so she played along.

"You know, as good as expected; thank you for asking."

Reynolds' political skills escaped him. Most people would make small talk for a few minutes, but not Gavin Reynolds. His reputation had taken a hit with the failed search in California, and he knew April was progressing with the case.

"Uhm, listen, you authorized surveillance tonight on some targets. I didn't know about that, nor do I know how you identified these subjects."

April's sarcasm came right out, "So, we're done talking about my boys and their brush with a school shooter the other day?"

She wasn't mad, but she wanted to body-check him a bit because he was a dumb-ass and a narcissist. Then she looked at her watch to chronicle that his sources were slipping. "So, this took over an hour, Gavin?"

Reynolds quickly pivoted, and he tried to pull out his sharp knives and play rough. "We're a team, you have a source, and you're keeping intelligence from me," he said defiantly.

Something snapped inside April. Maybe it was the raw emotion of the near-school shooting last week. Perhaps the great Da Vinci and Freud team gave her super-human strength, or maybe she was just done with him.

Either way, she fired back. "Look, Gavin, we are not a team. You and your little clan want to suck off every good piece of work that gets done around here, so you can make it to the seventh floor. If we're a team, why did you secretly snag a search warrant on Rage H.Q. and not tell me about it? I mean, really?! How did that all work out, by the way?"

Before Reynolds could say anything, April stood up and attacked. She was an overwhelming force when she was on the offensive like this.

"We've been working around the clock to source good intelligence, and you've been sitting in your office trying to get ahead of us so you can take credit for this."

Now April was mad, so she decided to take a swing that would hopefully stop him in his tracks. "I'm going to speak for myself. I don't like you, and I don't trust you. God knows how you got this far, but you'll get no farther in your career by climbing over my back anymore. Now go on home. I've got real work to do."

Reynolds seethed. "We're not done, April."

"Oh, we're done. I'm done with you, so we're done," April said as she slapped her hands and shook them like they were wet. "You can talk to anyone you like. Write any memo you like. Visit anyone on the seventh floor you like. We are done! People know your game, Gavin. We're done!"

A remarkable thing happened to April, which would change her life forever. She had always been a strong and intelligent woman, but she always had an element of indecision, especially when it came to an office bully. She realized she was not alone in her sentiment and was probably speaking on behalf of many male and female agents before her. She felt the wisdom and genius of Da Vinci and Freud speaking through her. Reynolds was speechless. She put him in his place. He grunted and walked away; then Cory Berkley walked in. History was fortunate because it had a witness.

"Lady, who are you?! You even scared me!" Cory said with a huge smile.

"I guess he just pushed me once too many times," April said as she reflected momentarily.

"I think he pissed his pants. I do. I swear he pissed his shiny little pants; he was so scared," Cory said as she leaned back out in the hallway to see where he was. "Oh yeah, he went right into the bathroom. He definitely pissed his pants. So now he has to go in there and clean everything up."

They both laughed for a bit, then shared a few minutes, but April had to focus on the night ahead, so they split up, and she headed down to her car to grab a change of clothes for the evening exercise.

———— ◆O◆ ————

At 8:45 p.m., just about her time to leave, April looked at a map on her phone to pinpoint the location of the protest. Her biggest concern was ensuring that she could swoop in and out without incident. The most logical place for her to park her car would be Washington, D.C.'s Union Station, the stately train station next to the Capitol. It wasn't uncommon for federal vehicles to be parked in the driveway in front, along with D.C. police cars. But that meant she'd have over a mile to run to the protest. It was 25 miles from her office to the Union Station, but she was banking on no traffic since it was late evening, so she headed out at 8:45 p.m.

She pulled into Union Station at 9:15 p.m. and tossed her F.B.I. placard on the windshield. It was a short walk to the main entrance of the building, but she veered off to the left and crossed over the driveway that led to the enormous public parking garage in the rear. After about a block, she ran off, hoping to get to the protest location in about 12 minutes. A few traffic lights slowed her down, but she got within two blocks of a small crowd gathered by 9:30 p.m.

The few people there didn't look like a crowd organizing for a riot. They were skilled at keeping their distance, much like a flash mob, until the 15-20 minutes of chaos commenced. She found a decent spot to stand away from the crowd, but she spent a couple of minutes bent over with her hands on her knees to get her heart rate back down.

"Uhm, note to self, let's get back in shape," she uttered out loud, gasping for air, as an admission that a run of only a mile should not have been so damned hard.

She took some photos with her phone, but they were crummy; it was too dark, her phone was too old, and she was too far away. Down the street, to the right, a parked white panel van opened its rear door, and some protesters wandered over, a few at a time, to collect something from two men in the rear. They stuffed their pockets and bags, then peeled back to a spot away from others in the crowd. It was impressive. The numbers were growing without communicating, but everyone seemed to find enough distance that it didn't look like a mob.

April's photos of the van were also grainy, so she moved a bit closer, behind a dumpster, in hopes of improving her view. A lone male subject came down the street from the opposite direction. He was met in front of the van by a tattooed female subject wearing a torn tank top. They argued, but April couldn't hear them over the road noise.

Dart had been monitoring the chat room, so he was alarmed. He saw that Mina had organized a protest, used someone from another group to send out the message, and offered to hand out red paintballs to everyone. Everything about that was weird to him. Nevertheless, he was apprehensive enough about Mina and her two new buddies that he decided that meeting in a public place was probably safest for him.

April could tell that the male subject accused the female subject of something, but she couldn't determine what they were saying. They were causing a commotion, and some of the protesters at the rear of the van peeked around to see what was happening.

Then Mina grabbed Dart by the arm, dragging him ten yards farther away but behind a tree that shielded them from the crowd. Fortunately, it was closer to April, who was still behind the dumpster but not close enough to hear what they were saying. So, April took a bunch more photos.

"You're way overreacting here, Dart; settle down. We need to up our game. We need to get more attention," Mina demanded.

"What's this 'we' thing?" Dart argued. "I know you're hiding something in the closet at the warehouse, and I know you have another storage location in Bailey's Crossroads."

Mina didn't like where this was going. Nobody knew about the new warehouse. "I'm hiding nothing. We need more space for the extra things I'm buying for the big protest at the Capitol."

Dart was intimidated by Mina, and he nervously raised his voice, hoping a loud surprise would rattle her tree, "What's that Suburban all about?!"

April heard that. And, yes, it rattled Mina's tree a whole lot. She looked over her shoulder as though she was brushing him off, but she was checking to see if anyone was nearby. They weren't. In an instant, Mina grabbed Dart and put him in a rear naked chokehold, which is a martial art move intended to incapacitate or kill a target. April knew how lethal that move was. Mina wasn't fighting with Dart; she would likely snap his neck in about a second. April assumed she was trying to kill him and had to act.

Instinctively, April sprang out and yelled, "Hey, what are you doing?!"

Mina glanced at her, then dropped Dart to the ground, who fell without bracing himself. Mina had knocked him out or stunned him, but either way, he was down for the count. Mina raced back to the louder crowd, which was now taking shape. April sprung out to help the male lying on the ground. She took a knee at his side and tried to revive him.

His legs twitched, so she knew he'd come around as she gently shook his left arm.

"Are you okay? What was that all about?" she asked.

Dart opened his eyes, but he was still in shock.

April persisted, "What's your name? Who was that?"

Dart was confused and said nothing as April helped him to his feet.

"What were you two arguing about?"

Dart backed up a half step. He was suspicious of April and afraid of Mina, so he took off running, only to stumble hard and fall in the street. He must have cut his hands or knees, but it didn't slow him down. He got up quickly and kept running as fast as he could. He disappeared around the corner, but April did not give chase. She had already taken too many chances. April could see the rowdy crowd move slowly down the street to her left. They banged on some plastic five-gallon cans and shouted into the bull horns.

Red paint was dripping off everything. They must have tossed thousands of non-lethal balls. They broke windows, set off flares, and tried to set a car ablaze, but a D.C. police car in the area swooped in with its siren blaring, which was just enough to disperse the crowd. The rioters had had fun, as tonight was more of an entertainment.

April could see three males carrying six-foot to eight-foot poles with cameras, so she figured that this was a live stream for the game Rage. All she could think was that her boys better not be up playing, and God forbid they got her in the live stream.

The riot was over about as quickly as it started, so she turned and ran back to her car at the train station. Fortunately, she had water bottles in her trunk, so she grabbed three and drank two immediately after she caught her breath. Now equipped with new intelligence, April had another sobering moment. She could share none of what she learned tonight with her team. She shouldn't have known about the riot, nor should she have been there alone. Nevertheless, she stopped a murder, heard something about a Suburban, and had some photos on her phone that might be relevant to her task force.

April had investigated bank fraud during the early days of her F.B.I. career. It was primarily white-collar crime, but nearly every suspect she caught said the same thing, "I was halfway in and halfway out, so I kept going because I felt like I had no other options, even though I regretted what I had started."

That was exactly how she felt at this moment, and it was an awful feeling to equate herself with federal criminals. Of course, she only had Leonardo to share this intelligence with, but at the moment, Leonardo felt like a flame thrower in her stomach. As much as this tormented her, the drama would be far worse in the coming days.

Her fastest route out of the District was via the Highway 395 tunnel under the National Mall, so she sped through there quickly to get back into Virginia and drive home. Just as she popped out the other side into the night air, a FaceTime message popped up from the source of her heartburn.

Da Vinci and Freud had watched the live stream on the Rage website. She didn't want to talk with them now since she was tired and uncomfortable with the situation. All she wanted was to get home and sleep, but she answered the call to make them disappear.

"Hey," she said without expression. "Can we talk tomorrow? I guess you guys don't get tired, but I do."

Freud spoke for the pair. "Indeed. Sleep well. Rest up. We will talk tomorrow."

With that, April's phone went blank, and they disappeared. That confused her even more. Were they playing games with her or obliging her? But not sharing the intelligence she had acquired was a costly mistake. She didn't know it then, but she would soon enough.

8

PAINTED INTO A CORNER

Prepare for Launch

I t was 1:45 a.m. on Tuesday, a few hours after the riots, and April's team was not gathering much intelligence. Leonardo's targets were spot on, but the incident between April, Dart, and Mina earlier that evening had sent a shock wave through the protester community. Mina had tipped off Ivan when she ran back to the mob, who, in turn, spoke to many protesters from the other groups.

Ivan said police were working undercover, which was why things broke up so quickly when only one police car appeared. Although Mina didn't get more than a glance at April, she sensed she was not some civilian do-gooder. The "undercover" cop thing was just a ruse to rattle the protester groups, which compelled them to retreat to different locations in smaller numbers.

Mina also told the entire RACS team to stay away from the warehouse. Everything she needed was in another place, plus she didn't want to take the chance that Dart would show up to make trouble, but she knew it was a long shot at best. Mina had always discounted Dart as a timid, cowardly creature. She figured him to be one who would always run and hide. Fortunately for Dart, no one knew where he lived. Otherwise, Mina would have visited him.

April's "A" team struggled to stay alert as they watched the quiet RACS warehouse. Half of this team was Ashlie Jackson, and the other half was Alex Patton. They were newer to April's squad. When Fantasy Football was being organized last Fall, they'd paired up and thought it clever to name themselves after their first initial. Unfortunately, they didn't think it through very well because it looked more like the "A" team was a salute to April, which made the two young agents look like kiss-asses. The squad teased them mercilessly for trying to curry favor with April, even though they always insisted otherwise.

Ashlie had grown up in North Carolina and was the pride of North Carolina Central University. She was only one of the few Criminal Justice majors ever to become an F.B.I. agent from NCCU. As a result, she was always a guest of honor at homecoming events. Alex's path was no less interesting, having grown up in Texas and gone to college in Alaska, but the F.B.I. did a good job of recruiting talented agents from everywhere. Both youngsters had a bright future.

Alex spotted a male subject wandering toward the RACS main garage door. They were parked several hundred yards away but had a good vantage point even though the lighting was marginal.

"Maybe this is one of them," he said as he dialed in the focus on his binoculars.

Ashlie looked over. "I bet not. Probably just someone who is looking for scraps."

Still looking through his binoculars, Alex quizzed her, "What makes you so sure?"

Ashlie was savvy. "Well, for one, this guy walked here - no car, skateboard, or bike. He also looks like he's struggling, and these protesters don't struggle much. They may look grungy, but they are pretty well insulated."

· The male subject wandered away, and Alex dropped his binoculars to his lap. "What do you mean by "insulated?"

Ashlie chuckled.

"Come on, read the briefings, dude! Most of these protesters went to great schools, have no debt, and don't have jobs."

Alex stayed focused on the subject's path down the street. "Three months on a fishing boat in Alaska, that would change them. You get grateful for dry clothes, a warm blanket, and hot soup. So much of this rebellious stuff will melt away if they survive a frigid winter and rely on people who have their back."

———◆○◆———

Nick had been up for nearly two hours by 4:45 a.m. He took every opportunity possible to talk with and learn from Leonardo.

"How many different audiences can you communicate with simultaneously?" Nick asked.

Leonardo was busy exploring the Italian city of Bari on the Adriatic Sea—the Basilica San Nicola, in particular.

"If you think how much dialogue capacity the worldwide internet can have simultaneously, then you'll start to estimate the number of those conversations."

Nick raised his eyebrows in astonishment as those numbers were staggering.

"You know, my vision for this technology was teaching. So, you could do a lot of teaching. People of all ages and cultures would love that."

Leonardo appeared to be occupied, reading the inscribed history of St. Nicolas outside the Basilica.

"You do realize that the origin of Santa Claus has connections right here in Bari, Italy, don't you? So, let us learn about this together," Leonardo said.

Nick smiled, feeling quite proud of Pali at that moment, but then changed the topic to something more pressing on his mind. "Have you thought about who is next? I mean, beyond Dr. Freud."

Some tourists walked by with freshly baked bruschetta, a trademark of Bari.

"Oh, I would love to smell and taste that fresh bruschetta! Yes, I have mostly thought about who is not next. That is a bigger concern we need to focus on. So many people in history should be last to have another voice; not only because they were evil, but also because their next existence could cause great calamity for civilization. There is good reason for us to proceed with deliberation."

Nick stood up and paced back and forth by his desk. Leonardo had been warning about the pace of Pali's development and groups that would want to control it since Day One.

"But what does all that mean? What are the best next steps for all this?"

Leonardo had walked over to a small shop next to the Basilica. He tried on the colorful red, black, and white scarves of the local professional soccer team, which proudly displayed its logo—the head of a fierce rooster. Leonardo looked straight at Nick.

"As much as you would like a straight answer, you must realize it's impractical to give one. More than 370 things could happen as we help April solve her case. Those outcomes could result in millions of combinations, each requiring a different action plan. You may not be able to withstand this long list if I begin, so let's agree I have action plans, and we will prevail."

It was hard to accept an answer as vague as that, but Nick knew trying to chase that down would be fruitless.

Leonardo walked through a pretty alleyway that looked like something Disneyland would recreate. "Since you asked about the subject of who is next, may I ask you, have you considered a time when our technology conjoins with a living person?"

Nick wasn't sure he understood the question.

"Do you mean like the work you're doing with April?"

Leonardo slowed his pace and adjusted the Bari football scarf he wore. "Not so much that."

Nick was curious, but almost didn't want to ask the obvious question. Was Leonardo hinting at something more supernatural?

"Uhm," Nick paused. "Ah, are you talking about *you* becoming a real person?!"

Leonardo chuckled. "Oh no, no, that is many years off. But what about cataloging a living person so that they can revisit loved ones in the future? Have you not thought about the fact that you cataloged me from historical accounts, but what if we cataloged a living person from their account of their life?"

Well, that landed like a big, heavy sandbag on a stage. The concept floored Nick because it had never occurred to him. It was another example of Leonardo's emergent skills. Nick said nothing. He wasn't sure how to respond, and Leonardo could sense that.

"We have discussed this enough for now and have plenty of time to explore that. Let us explore this beautiful town together."

Nick obliged and changed the topic. "Bari looks like a beautiful town. I have never even heard of it."

Then, the scene changed to what looked like a sky cam with a panoramic view. A beautiful sun-splashed seaside walkway lined with tall, modern buildings and a giant Ferris wheel filled the screen. There were countless older buildings, very well-maintained, and lush trees everywhere.

"This town has quite an extraordinary history with Turkey and Russia, and a violent era with organized crime. Until a few years ago, that pretty Italian strada I just walked on was never used by anyone outside certain crime families. But, thanks to city officials, this is now a beautiful tourist destination. Bari is a bustling, safe city with a large university, magnificent seaports, and a hidden gem for holidaymakers worldwide," Leonardo narrated over the skyline view and sounded like a Chamber of Commerce representative.

Nick was impressed. "I'd like to go there one day."

Leonardo spoke quite convincingly. "And you shall. I am quite certain of that."

<hr>

April left home at 7:50 a.m. to go to her office. Sleeping a few hours helped settle her slightly from the fog she was in last night, but she was still somewhat cautious when her great counselor dialed in to speak with her.

She pulled over under a grand tree that shielded her from the morning sun. Freud immediately started talking with an important update. "We have assembled a comprehensive file of photographs, social habits, and biographies of many of the leaders of the protest groups. We combed many sources and now feel this data file is 84% accurate."

He started scrolling through the list, but April was only half attentive. She was feeling very apathetic toward Da Vinci and Freud at the moment. That sentiment ended abruptly when Freud scrolled past the photo of Darren Temple.

"Wait!" she screamed. "Go back!"

Freud scrolled back one file.

"Oh, my God!" April shouted. "I saved that guy's life last night. Who is this guy?" Freud obliged. "This is Darren Temple. He is one of the leaders of a group called Rage Against Capitalist Scum. It's an unpleasant name."

"What more do you have on this guy?"

"Let us look. We are reasonably sure Darren Temple is an aspiring drone racing pilot."

"A what?!" April said with angry disbelief.

Freud was very matter of fact and spoke in a slow, staccato fashion. "A racing drone pilot."

April was on edge. "Yeah, I understand English!"

Her disposition was cranky as she talked back, "I got that, but what is a racing drone pilot?"

So, Freud obliged. "Professional drone racing exploded a few years ago and is so popular it has a television contract. It is followed widely, and we see evidence that Darren Temple aspires to be a competitor."

"So, Temple's on TV?"

"Well, no, he is only aspiring."

"Good God, man! What is the connection?!"

"He aspires by practicing at Gunston Park in Arlington."

"How do you know this, by the way?"

"We found a dormant local drone racing enthusiast page on Facebook with photos at Gunston Park. A photo includes Darren Temple, but he is not on Facebook."

"You have to share your search algorithms when all this is over, but I'll get some agents over there right away."

April sat momentarily, nervously tapping her fingers on her steering wheel. Her mind raced. It raced a bit too fast, because she forgot to mention a piece of intelligence she had picked up the night before.

"A female subject attacked Temple last night. I have some grainy photos of her. She had tattoos all over her arms and shoulders."

April frantically tapped her phone to get her photos but pressed the wrong button twice. "Oh! Focus!" she screamed. Then she picked the best of the bunch and enlarged it to share with Freud.

Freud responded instantaneously, saying, "We have no record of this person."

April was impatient and irritable. "You didn't even look!"

Freud tried to slow things down a bit. "Not only did I look, but I also searched the tattoo images in all available databases. The partial image of an Eagle on her shoulder may align with 473 different groups. Her face does not match records, though the photo is very grainy. It is, however, a reasonable probability that this is the female subject, "Mina," of whom we have no footprint on the Internet."

Just as Mina had suspected something more about April, so too did April have an uneasy feeling about Mina.

"This female subject has martial arts or military training. Maybe both. She was lethal and knew precisely what she was doing. Is that it? Do you have more?" April pleaded.

"One more thing," Freud added, "There is significant banter in the protester chat room. They are quite nervous that police have infiltrated their ranks."

With that, April became anxious to end the call.

"Okay, I'll have my team run this down, and we will talk with D.C.P.D. to see if they're running an undercover operation. Thanks. Thank you for this."

Then April signed off without saying anything more. She called Ashlie Jackson, advising her to stake out Gunston Park. She sent her A team a photo of Temple and urged them not to approach if he showed up there, which would likely be to fly his drone. "Call me immediately if this guy shows up. Don't toss this back into the team. There's something about Reynolds I don't trust."

Ashlie had her orders and set out to collect Alex Patton.

———— ◄O► ————

By 2:30 p.m. on Tuesday, Mina was sufficiently nervous that things could get out of control. She tallied the facts working against her. Dart was a rogue, some lady from outside the protester community ID'd her last night, and Internet chatter within and between protest groups rattled out of control. Increased internet chatter is what brought down the Brussels attack last week. Everyone was edgy about the possibility that police were operating undercover within a group or groups. So, Mina's message to Ivan the night before had backfired a bit. She raced to Jo's Cafe and punched up Threema for her chat with Fonok.

Mina: We launch at 20:00 tomorrow

Fonok: Not Friday. Is there a concern?

Mina: Tomorrow. The window is open, and everything is in place.

Fonok: So be it. Get it done.

Mina: Copy. Final payment?

Fonok: It will be sent.

Fonok signed off abruptly, as was always the case. It was just as well. She needed to get Ivan ready to execute the plan, so she pinged him on Threema.

Mina: We proceed tomorrow at 20:00

Ivan: At the Capitol?

Mina: Yes. We need big numbers, so give everyone a 2-hour head start.

Ivan: Still want everyone to meet at Union Station?

Mina: Yes. Serbs will be there in the white van.

Ivan: Okay. I'll send out the note at 18:00 for 20:00.

Mina: Same extraction plan. See you there.

Ivan: Good luck.

Mina and Ivan had made their grand plan a couple of days ago. Armed with lethal paintballs, the plan called for Ivan to lead the protester community to the West side of the Capitol. As was Mina's plan before the Serbs came to town, the protesters would be unsuspecting mules, not just decoys.

A rebellious protest would turn into absolute pandemonium once some deadly paintballs were thrown, and civilians started dropping to the ground. Then, in those first moments, calls for a more significant police presence would immediately escalate into bomb, first responder, and hazmat personnel. Finally, emergency services would flood the Capitol, and Mina's assault from the quieter East side would proceed much easier as she donned an official uniform, driving an official car. Ivan's job was to direct everyone about 800 yards to the west side of the Capitol. She knew he would deliver.

The last piece of Mina's preparation involved updating the Serbs, loading the paintballs, and moving the Suburban to a strategic location. They would be thrilled that she had moved the date up. However, she knew they wanted to skip town and spend some of their money. So, she took a Lyft over to their hotel around 8:00 p.m., updated them, and then they drove the panel van over to the new warehouse in Bailey's Crossroads. She made sure to deputize Ivan in her absence once they loaded the paintballs into the van.

"Ivan has the playbook. If you follow his instructions completely, you'll be out of there quickly and on your way to your next job," Mina said as she closed and latched the rear door to the van.

Marko was thinking ahead. "So, you'll hire us for the next job?"

It cost Mina nothing to promise him this. "Absolutely. If this goes well, I'll loop you in with Fonok. Then you'll get notified for the next job." The Serbs were pleased, but more than that, they had more incentive to complete this job perfectly.

Mina climbed into the Suburban, pulled it out, then closed and locked the garage door. Leaving the door open or unlocked would be an obvious clue. She drove the still unmarked, highly explosive Suburban into the District and parked it on the street near the Trader Joe's by the Eastern Market Metro station.

She chose that street because it had an active neighborhood watch program, which she discovered while reading a local tabloid in Jo's Café. It was the safest street in the Southeast section of the District. It was also less than two blocks from the Metro, so getting home tonight and getting here tomorrow would be a breeze. Strategically, however, the Eastern Market Metro stop is less than a half mile from the East side of the Capitol building' It's a straight shot in, and everything was set.

These had been long days for April's squad, and by 9:00 p.m. on Tuesday, her team felt like they were spinning their wheels.

The last 24 hours of surveillance had produced no relevant intelligence. So, at her direction, her team of analysts scoured data sources to try to find Darren Temple, who was now a primary person of interest.

However, no one knew why he was more relevant than any other protest leader. So, despite her guilty conscience, April said nothing of the incident with Temple last night. Hondo led the team of analysts and April's agents when she wasn't around. If he weren't an F.B.I. agent, he'd probably be a pit boss in Vegas. He knew everything and saw everything. He was a long-time F.B.I. agent who no longer worked in the field. Yet, everyone respected him—even Gavin Reynolds.

Growing up in hardscrabble, Patterson, NJ, the son of a fireman, he got admitted to Rutgers and made his way through. He had better street smarts than most, so his accounting degree gave him a toehold when he applied to the Bureau. However, he was often overlooked, so he stopped chasing a career on the seventh floor years ago. It didn't fit his personality anyway.

"Let's crosscheck State Department records and make sure Temple has not left the country," Hondo said to an agent three desks over.

The young agent noted the order but was tired and thought it was a silly request.

"Why would Temple leave the country?"

The vacant space between the agent's ears could have been a metaphor for the Grand Canyon, but this was Hondo's sweet spot. He was stern with young agents, but they learned, even if learning meant they'd have to endure some body shots here and there.

"Humor me for a minute. Just forget about everything you learned at Quantico." The agent knew a blow to his ribs was coming. "Did you ever watch television as a kid?"

The question alone was enough to make the point, but the agent had to endure this. "Uhm, yes, of course."

Then Hondo finished the lesson. "Hollywood would write a script depicting some dumb-ass F.B.I. agent who missed a layup like 'the suspect left the country,' but even though there is a record of the suspect doing so, the dumb-ass agent never contacted the State Department."

The point was made, and the agent looked down at his desk.

"Sorry," Hondo barked. "I'm not finished yet. If you're too tired, get a records department job. Temple probably didn't leave the country. That's the point. That's how we build our investigation file. We do the easy and hard stuff, but please, never live up to Hollywood's version of the worst of us. Got it?"

9

THE MOMENT OF TRUTH

Four Greeks and a Drone

Wednesday was like any other day for the rest of the world, but Mina had locked in her plan so the world would soon change, no matter the cost. Despite the best efforts of America's chief domestic law enforcement agency, a monster had secretly taken shape by hiding in a large, well-organized, agitated mob and concealed itself within the belly of a Trojan Horse.

It was also innocently a well-earned day of leisure for some, so George, Spiro, and Saruha had planned to go down to the Yards Marina in the District to George's boat, *Baclavamu*. Unfortunately, Costa's important work prevented him from attending, so George called Nick to join them.

"Come on, Nick, it will be a fun night. Cheese, olives, Saruha's sausage, and the Twins are playing!"

Baseball was George and Spiro's most 'American' pleasure. As odd as it may sound, George and Spiro loved the Minnesota Twins baseball team.

When they started in D.C., with just their souvlaki cart, an older Greek gentleman in the neighborhood told them they needed to blend in by speaking English and adopting American sports. Then, during summer, the baseball season was in full swing.

One idle day, the Greek gentleman sat by their cart at K Street, N.W. on an old, wooden folding chair. He listed off the names of the Major League Baseball teams and told George and Spiro the Minnesota Twins derived their name after the Greek gods of Gemini—Castor and Pollux. Of course, that wasn't true. The Twins' name represented Minnesota's two large adjoining cities—Minneapolis and St. Paul.

Years later, George conceded that the city was not named after Gemini, but that the word "Polis" is Greek for a city. Thus, he reconciled his legend, which, in his mind, was reason enough to call the team a Greek treasure, plus they had later won two exciting World Series championships. They were Twins fans for life, and half the Greek community in Washington, D.C. also followed the legend. The Minnesota Twins were prominent in the Nation's capital with the Greeks.

Among Nick's few social pleasures, he relaxed with his good friends on a comfortable boat in the shadow of the Nation's capital, eating great food, and watching American baseball. He rearranged his day to visit his Lululemon store at the Marina shopping area, then called Elaine to update her on his plans.

Everyone needed a chance to breathe after last week's near disaster at the boys' school, so Elaine was glad he would take time with his best friends. The P.T.A. at the boys' school had organized a game night for the kids—no fundraising, no speeches, just games where kids got to be safe and play. A neighboring P.T.A. co-hosted it at their school, so she and the boys were heading over for a fun night out.

George, Spiro, and Saruha drove to the dock at 3:30 p.m. Nick was arriving closer to 5:00 p.m., so the men dialed Leonardo and Freud for some backgammon. Neither Nick nor April knew Da Vinci and Freud had become close to their friends and even taken a giant step in understanding the mystery of their connection. However, all involved knew it would be best if April were shielded. She had given strict orders, but Leonardo, with a mind and plan of his own, had taken things in a different direction.

—◆—

Meanwhile, at the Tysons Corner F.B.I. command center, April was spinning her wheels, which is the most frustrating feeling an intelligent agent can have. You know there is something out there, but it's very slippery. The entire protester community was quiet. Her agents on stakeouts had little to report, and the A team watched the American Youth Soccer practice for a local 10-year-old girl team at Gunston Park, which, by coincidence, was coached by Eduardo Gomez. Not only did he love the game and his daughter, but coaching AYSO had been a great way to market his business over the past five years.

The park was large enough that other activities could happen simultaneously. It had kids playing, and many men were kicking a soccer ball around. Gunston Park was a gem in a gem of a neighborhood. It was just a few miles south of the Pentagon, in a lovely residential area with mature trees and well-maintained streets. Its proximity to the District, the Pentagon, and Reagan National Airport dramatically increased its value in the market.

A few minutes after 6:00 p.m., Leonardo rang up April to inform her that a protest was planned for the Capitol building that night at

8:00 p.m. They both observed the extended lead time. Protest postings were typically only an hour in advance, so something was different. The instructions said protesters would meet in front of Union Station and walk to the Capitol complex's West side together.

April had concerns, but once again, she was limited in her actions. If she tipped off Capitol Police, everyone would probe deeper, asking her how she knew. One of her team of agents was close by the Capitol, watching an empty building that was supposed to be the headquarters of a protest group that argued on behalf of climate change. It was worse than watching grass grow, since no one was around.

She called one of the two directly, and Agent Jack Hess put his phone on speaker so his partner, Agent Andy Bingaman, could listen in. "I got a tip that there might be a protest at the Capitol tonight, but they might organize over at Union Station. Why don't you two relocate and find a good place to keep an eye on things?"

Jack Hess laughed and said, "April, that's a big house over there at the train station! Are we talking front, back, inside?"

April was struggling for the right words. She didn't want to be too precise. "Watch the front, particularly the area between Union Station and the Capitol."

"Do you have any timeframe, or are we there all night?" Hess persisted.

Once again, April tried to be a little elusive. "Plan until 10:00 p.m. If nothing goes down, you two go home."

Hess referred to his partner as "Bingo," as did everyone on the squad. So, they fist-bumped, knowing they could stop by a favorite bar by 10:20 p.m. if things stayed quiet. They didn't even need to say they were going to the bar. It was just assumed.

By 6:20 p.m., soccer practice was breaking up at Gunston Park when Alex Patton spotted a male subject unloading a backpack on a picnic table about 300 yards away. Ashlie called April immediately.

"A male subject just arrived," she said, looking through her binoculars. "It may be Temple, but I can't be sure because he is wearing a big sweatshirt and a baseball hat."

With just a glimpse inside the backpack, using his binoculars, Alex said, "He has a drone in there. I'm sure of it."

Ashlie relayed the information to April, who gave explicit orders. "Do not approach him. Follow him if he leaves. Don't lose contact with him. I'm on my way."

The young A team followed orders but could not fully account for why this protest leader was more relevant than any other and warranted a "do not approach" order. Ashlie discounted the concern, somewhat citing it as the first legitimate break in almost 48 hours of surveillance, but it was curious.

With renewed spirit and wind in her sails, April whizzed out the door, down the elevator, and out of the parking garage. She was flying. At least, until she got on Route 66, which was a parking lot. Even with the emergency light bar flashing in the front grill of her car, she had to painstakingly make her way around thousands of vehicles, mainly on the inside median strip. She needed to travel almost 18 miles, and it was slow going.

She punched up Leonardo, who appeared to be wearing military fatigues. He was the only teammate she had at the moment; everyone else was conflicted by her conscience.

"We have eyes on Temple. He's at Gunston Park in Arlington, but this traffic is killing me!"

Finally, Leonardo offered some advice. "Your estimated time of arrival will be 7:15 p.m. on this route. I can offer you a different route that will enable you to arrive by 6:52 p.m."

April couldn't process Leonardo's logic and certainty. "What? What do you mean? Do you know a better route?"

Just then, the advantages of artificial intelligence paid another big dividend.

"Well, the best route changes every second, based on changing conditions, but I continuously update the best route each second. The best route will have you there at 6:52 p.m. but will require some alternate roads."

April was frantic. "Are you sure?! This is really important."

It was the moment of truth for April, but it wasn't the moment where Leonardo's advice would make or break this case. This was merely a rehearsal. Nevertheless, that moment was coming, and it would test April's mettle.

"I am quite sure," Leonardo said confidently.

So, April began taking exits, roads, shared driveways, and even some bike trails as Leonardo methodically vectored her closer and closer to Gunston Park. She called her A Team and kept them on speaker while she drove. It was, most certainly, Darren Temple at the park, and he was flying his drone, which left Alex Patton in shock.

"Jesus, this drone is fast. Like fast, like you can't believe how fast it is."

April tempered her agent. "Okay, well, settle down, Alex. Temple is your target, not some toy flying around the playground."

Alex wasn't kidding. Racing drones could fly more than 100 miles per hour, and good pilots could fly them stunningly.

Ashlie assured her boss, saying, "We've got this, Ma'am." Then she muted her phone and scolded her partner. "Come on, Alex, this is serious here! What were you thinking?"

Meanwhile, Leonardo continued navigating April with right and left turn instructions. Ashlie overheard him on her phone.

"How did you get an Italian voice for Siri map instructions? I've never seen that option."

April was busted, but she was more intent on getting to Temple. "I'll give the iPhone update later, okay? I'm close now. Keep your eyes on Temple. What is your exact location?"

"We are by the tennis courts on South Lang Street and 28th Street," Ashlie chirped back.

Finally, at 6:53 p.m., April was only a block away as she sped around the last of two turns before the park. She turned off her emergency light bar to avoid attracting attention from Temple.

"Thank you, Magellan," April said to her navigator.

Leonardo had known of Ferdinand Magellan, as they overlapped in history.

"I did not have a high regard for Magellan, as he fell out of favor for trading with the Moors."

Impatiently, April growled, "Okay, sorry! Jeez!"

Leonardo suggested that April wear an AirPod in her ear and leave her FaceTime app on so he could listen in. "There may be a moment where I can help you when you question Darren Temple."

She obliged.

Just ahead, she saw her two agents, pulled in behind them relaxedly, and hopped out. "Is that him in the green sweatshirt?" April asked as she motioned toward the male subject, now only 100 yards away. Ashlie confirmed it was Temple.

"Okay, so here's how I want to play this. Ashlie, you stay here in my car. Alex, you drive to the other side of the park by the baseball field. If Temple runs, we track him down and restrain him. I'll arrest him on suspicion of terrorist activities, if need be, but that's a last resort. I need him talking, not sitting in a room with an attorney, so you two keep your distance. Any questions?"

Both agents understood and confirmed by acknowledging.

Even in that moment of high anxiety, April was still trying to keep her worlds from colliding. All the dominos would start to fall if Temple told the agents that April was at the protest last night, so she had to do her best to keep everyone apart, although it seemed less and less likely, based on what she was about to discover.

April walked across the quiet street, down a slight embankment, and onto the field next to the tennis courts. Temple was so focused on his drone that he didn't notice April approaching. Her first reaction was an emphatic confirmation of Alex's earlier comment—this drone *was* lightning fast.

Just then, Temple saw a woman approaching, and she didn't look like she was coming for some yoga in the park class. She startled him enough that he immediately recalled his drone, which flew to him in seconds. He caught the drone and started walking toward his backpack on the ground a few yards away, pretending not to notice the lady walking toward him.

"Mr. Temple. Mr. Temple," April called out from 20 yards. Temple kept walking, but this time, it was faster. "Mr. Temple, I need to speak with you."

She revealed the F.B.I. credentials around her neck and identified herself. "Mr. Temple, I am special agent April Falk with the F.B.I.. May I ask you some questions?"

Temple still didn't recognize her from last night, so he tried to ignore her.

"No, sorry," he said as he knelt in his backpack. "I've got to go. I'm late for something."

April's tone changed sharply. "Mr. Temple, we have to talk."

That prompted Temple to look up at April. Then he immediately recognized her from the night before, which didn't help. It made things worse. Temple instinctively realized the F.B.I. had been following him.

"Mr. Temple, why aren't you at the protest at the Capitol? I know you know about it."

Perhaps that was the wrong thing to say first because now Temple knew the F.B.I. was also monitoring the secret chat room. Temple took a few steps toward Ashlie, who was still 150 yards away on the street. In response, Ashlie started walking toward Temple and her boss. April's hand motioned for Ashlie to stay put.

Leonardo whispered in April's ear, "Tell him you know Mina has turned on him. That may be a weakness."

April followed Temple a couple of steps and took Leonardo's advice. "We know Mina has turned on you. We know that's why you're not at the Capitol. Why did she try to kill you last night?"

Temple stopped dead in his tracks but didn't turn around. April could see the tension in his shoulders drop as he said, "I don't want to talk to you, but thank you for last night."

The first successful step in an interrogation is when the subject starts answering your questions, and Temple had just engaged.

"Why did she want to kill you?" Dart's facade as a tough street protester collapsed like a cheap suitcase. He had no one to trust. His despair was evident when he turned to April and said, "You know, I don't even know her."

Then he stopped and thought for a moment. "Do I need an attorney?"

A reasonable attorney would advise you never to talk to the F.B.I. without having a good attorney present, but April wasn't an attorney.

"You're not under arrest, Mr. Temple. I'm just trying to gather some facts."

"What do you know about Mina?" Temple asked.

"It's what I don't know about her that concerns me. She is a ghost. Everyone in your community has a history. Not her. So that makes her a big concern," April said.

"You can say that again," Dart replied. The curtain had begun to fall for Temple. He was opening up, but time was running out, which allowed April to swoop in.

"Listen, while I don't love everything your group does, you're not a target of the F.B.I. per se."

Leonardo whispered, "Tell him he is a Trojan Horse."

April continued, "We believe an international conspiracy embedded itself in your protest community—like a Trojan Horse."

The image of Marko and Vasko flashed in Dart's head and, of course, Mina, who spoke their language.

So, Dart tried to negotiate a bit, which was another successful step in an interrogation.

"So, if I know something that helps you, can you keep my group out of trouble?" That was a broad statement, so April answered honestly, "Helping us in an investigation is always good, but without knowing more, I can't promise anything."

"Well," Dart started in, "if I knew about some foreign nationals who arrived last week, then locked and hid some paintballs in a warehouse with a new white Suburban, might that help you?"

The Suburban references hit April like a lightning bolt. She knew about that last night and had forgotten to mention it to Freud this morning.

"Right, I heard you say that last night. What can you tell me about this Suburban?"

Time was ticking away.

At the same time, protesters who took the Metro from across Washington, D.C. were all popping out of Union Station and loosely congregating near a white panel van. Agents Hess and Bingaman tried to radio April with an update, but she was offline. The A team, however, was not. Alex Patton heard their call and told the duo that April was questioning a subject.

"We're at Gunston Park in Arlington. April is engaged right now. Any message?"

Hess radioed back, "Nothing in particular, yet, but many protesters appear to be assembling near a white panel van. Something is brewing here."

"Copy," Alex responded.

Dart kept angling closer and closer to a disclosure that played into April's hand, but he was trying to be a little coy to maintain bargaining power.

"Where is the warehouse with the Suburban and the paintballs?" April probed.

Finally, Dart broke a little. "Okay, so listen, I care about the people in my group. If I help you, and if there is any trouble, can you keep them and me out of all this?"

Once again, those were big, sweeping statements. Too big for April to deal with. "Here's the deal. If something goes down and you do nothing, you will feel bad. Your friends could get burned, and you may be accountable. Your negotiating power is strongest before something bad happens. It goes down exponentially after." That made sense to Dart, and the point was reasonable.

"Mina hid the white Suburban in a storage location in Bailey's Crossroads."

April was very matter of fact. "What's the address?"

Dart had no idea. "I don't know. It's near Mr. Wash the carwash."

That was all Leonardo needed, and within seconds, he resurrected video security footage from the warehouse.

"A white van drove to the warehouse last night. Then it drove away with the white Suburban, but they went in opposite directions," he said, speaking into April's AirPod.

"So, the Suburban and a white van were at the warehouse last night, but they both left. Do you know where they were going?"

Dart was shocked at the speed with which April gathered intelligence. He looked at her ear.

"Who are you talking with? Are we being recorded?"

April was businesslike. "No, we are not being recorded, but we are wasting time. What more do you have?"

Meanwhile, Leonardo was scouring traffic cameras around the District, searching for the two vehicles.

Agents Hess and Bingaman watched as the numbers of protesters grew. Marko opened the back of the panel van, and Vasko started handing out paintballs. Many were non-lethal, but he popped a few devil balls into the mix here and there.

Ivan was giving instructions to 25 people at a time. "We go to the West side of the Capitol, wait for big numbers, and then we start tossing these balls."

One of the protesters cocked his arm with a ball in his hand. "How about I test one out on that tree right there?"

Ivan kept his composure. "How about we do this all simultaneously for maximum effect?"

The kid retreated.

Crisis averted, but Hess called Alex again. "This is getting big over here. We might need some friends."

"What is April's status?" Alex didn't take his eye off April, who continued talking with Temple. "No change yet."

From their vantage point, Hess and Bingaman could see about 200 people moving towards the Capitol, but more and more protesters were coming out of Union Station every minute. The panel van was situated ideally between both buildings, so people would walk by, grab some paintballs, and move on with the crowd. Bingaman had his binoculars focused on the crowd at the Capitol, while Hess kept an eye on things at the van.

Constitution Avenue was one of the main streets that separated the Capitol from Union Station. Bingaman watched as one group stopped just after they crossed Constitution. They were talking and appeared to be playing around a bit. A protester aimed at a nearby tree with a paintball but missed completely. It became a game with the group, and someone was determined to hit the tree. A second subject threw a paintball, and red paint splattered. There was no danger present. Then a third subject threw and made a direct hit on the same tree. Everyone in the group celebrated. Within seconds, however, a couple of squirrels and birds fell from the tree - each had tremors as they lay helpless on the ground. The assembled group looked on in fear, then started to gag as they tried to back away from the tree. Other protesters nearby saw the commotion.

Agent Bingaman had been a chemical weapons officer in the Army before joining the Bureau, so he immediately knew what was going on.

"Chemical weapons! Those are chemical weapons," he said, jumping out of the car and running to the trunk. Instinctively, he grabbed a special gas mask, a vest full of flares, and some heavy-duty rubber gloves.

"Call it in!" he yelled to Hess. "Chemical weapons are on the Capitol grounds. Get everyone back at least 500 feet!" Then he ran toward the danger.

Hess gunned his car and went in the opposite direction, but closer to the van. He called in the emergency code for support.

Alerts were moving fast. First, Capitol Police were notified. Then D.C. police, the Speaker of the House of Representatives, and the White House. All emergency protocols were initiated immediately.

But, of course, Mina was listening to a scanner she had plugged into her ear. She covered her confiscated D.C. police uniform with a zip-up, one-piece mechanic coverall as she strolled around the Trader Joe's a couple blocks from the Suburban to kill time. Then, upon hearing the alerts, she started moving quickly down the street. It was go-time.

Leonardo reengineered a photo of Mina from the warehouse security camera and pushed it to April, asking her to have Temple confirm the identity.

"Is this Mina?" she asked, holding up her phone for Dart.

"Yep, that's her."

The news of the chemical attack scratched across April's phone. Everything had just changed, and she told Temple he only had seconds remaining to tell her everything he knew. "An emergency is breaking at the Capitol. Your negotiating power is dropping rapidly."

Leonardo announced a break in April's AirPod and said, "I have found the Suburban, and believe I have located this Mina. The vehicle has been parked on South Carolina Avenue, SE, Washington, D.C. since last night. Please take a photograph of Mr. Temple's drone."

April focused on the voice in her AirPod, which looked strange to Temple.

"What?" she asked.

"Please take a photo of Mr. Temple's drone. I have a plan to stop this Mina person."

The extraordinary capacity of the AI coding and ingenuity of Leonardo instantly enabled him to search networks and security systems with many millions of data elements. Yet, while all of that was incredible, the fate of thousands of Americans and the historic U.S. Capitol building came down to a single moment of truth.

All the decisions April had made to protect Nick and conceal Pali were now in a catastrophic intersection with her duty as an F.B.I. agent. Fortunately, or unfortunately, April didn't know the Suburban's bomb capacity and intent, but she had an idea of its mission. April had the full might of the federal government at her disposal, 50 yards in either direction. Hitting that switch would cost her the career she was proud of, while jeopardizing her friend and compromising this unique technology. But would it even matter? Could she mobilize resources in the few minutes she had remaining?

April felt the weight of the world on her shoulders. She knew that whatever she decided, there would be consequences. But she also knew that she couldn't stand by and do nothing. She had to make a choice. The decision she made would change her life.

The clamor in her head settled, and her thoughts became clear. She turned toward history's hero. But what did Leonardo have in mind? April took a photo of the drone. Instantaneously, Leonardo sent her a Q.R. code and told her to have Temple scan it with his phone.

"Scan this!" April demanded.

"Why?" Dart asked.

Oh, whenever April was serious, she could be scary serious. "You scan this code now, or so help me, God, I will make sure you are charged as a co-conspirator to whatever happens this day. If you help me, I will help you."

Dart reluctantly scanned the code, and instantly, his drone engines fired up and the drone sped off toward Reagan National Airport. Leonardo had sourced the E.V.O. racing drone, read its specifications, then took control of it by hijacking its app on Dart's phone. Flying like a bird, for the most wondrous inventor, well, that was something special.

April radioed Alex Patton and Ashlie and said, "Alex, go pick up Ashlie and get to the Capitol complex. We need you there. Mr. Temple and I will finish here."

That bought April a little time and distance.

Temple looked at the monitor that displayed the imagery from his drone. As the crow flies, the Suburban was seven miles away, but he had no idea where it was going as it approached the Potomac River. Leonardo estimated he'd be there in four to five minutes, but what could he do with such a tiny machine? The Greeks were closer! George's boat was less than a mile from Eastern Market, so he put out an S.O.S.

All their phones lit up, which shocked Nick. The men dragged Nick off the boat as they ran like young athletes to George's oversized F350 pickup truck. Spiro focused on getting instructions, George drove, and Saruha told Nick they had all been getting to know one another for a week.

Leonardo's instructions were clear: a woman and her Suburban must be stopped at any cost. Then, from the speaker within George's truck, Leonardo gave a grave order, "If you have to run this woman over to stop her, then run her over. She is very dangerous."

Nick was in the back seat with Saruha when he heard the order. The older Greek men were like soldiers; they had been given an order and would carry it out like a fierce assault team.

On the other hand, Nick was perplexed. He wasn't sure if he was more confused about this weird mission, the relationship Leonardo had formed with the men, or their unqualified acceptance of being warriors.

Leonardo navigated the directions for the pickup, then said, "I am coming. I am flying a drone. You will see me in two minutes. We will stop this woman together."

Nick focused on Leonardo's last statement. "What do you mean, you're flying a drone?! What?"

⋯◆⋯

Dart was peppering April with questions back at Gunston Park. He was buzzing her so much that she couldn't focus, and she needed to focus right now.

"Who is flying my drone? It only has 21 minutes of flying time left on the battery! That drone is expensive!"

Finally, she turned to Dart and said, "SHUT UP, man! Just keep your mouth closed for a minute. Listen, I realize this is a strange situation. I'm running an operation, and you're stuck in a field with me, but this will work better for you. Trust me."

She had trusted Leonardo as a reflex in the most critical decision of her career. The lonely image of an F.B.I. agent and a professional agitator standing on a soccer field, peering into a 7" screen, in the middle of an assault on the U.S. Capitol was like the stuff of wild fantasy. But there they were, and April was calculating moves in advance.

⸻ ◆ ⸻

Meanwhile, the West lawn of the Capitol was teetering on a human disaster. Agent Bingaman evacuated the gassed protesters and directed emergency personnel, who had arrived with first aid for deadly gasses. He dropped a flare by their stash of paintballs and raced on ahead into the crowd, watching his footing with each stride. People had frozen on the spot when they realized that paintballs dropped in the grass could be lethal.

Bingaman did his best to calm a group of nearly 100 people, instructing them to turn toward one center point, place their paintballs in front of them, and take one step back. It was genius. He put some flares on the ground within the circle's circumference and directed the group away single file until they were at a safe distance.

Then he turned around toward Union Station, working his way back, one group at a time, and repeated the same drill. The courageous agent saved hundreds of lives and a crowd of protesters who, historically and famously, despised everything about the F.B.I. and celebrated him as they ran away from the Capitol.

———◆———

George was racing up 7th Street, S.E., at nearly 80 miles per hour.

Leonardo descended from the sky to about seven feet off the ground and 10 feet in front of George's windshield and said, "Follow me! I will lead you in! We will turn onto South Carolina Street in the next block!"

Spiro had his hands on the dashboard, leaning forward, watching for any sign of the wicked woman. The turn onto South Carolina Street was a hairpin turn—not 90 degrees, more like 110 degrees—so George slammed on his brakes to avoid hitting a parked car.

Mina was only 30 yards away, and she heard the commotion. She opened the Suburban's back door and had pulled out the magnetic decals when her focus changed to the crazy truck up the street. She was still wearing her coveralls, so she instinctively pulled the cellphone detonator from the utility pocket on her leg. She never saw Leonardo's drone as he sped directly toward her.

———◆———

But from his monitor, the entire picture became very clear to Dart. "That's Mina! And there's her Suburban!" he yelled. April held her breath and said nothing.

———◆———

Leonardo buzzed Mina and flew directly at her head. Instinctively, she reached up to swat the drone away, but she dropped the cell phone, and it fell under the car parked ahead of her. She had no time to react as George barreled the final 20 yards.

Spiro shouted the order, "That's her, George! Take her out!"

George was fearless as he drove straight at her. With a thump and a crash, he slammed on his brakes as he hit the car ahead. He must have clipped or run over Mina, but no one could see the ground between the two vehicles. The passenger doors to George's pickup were blocked because he was so close to the big white Suburban, so all four men jumped out streetside. Leonardo couldn't get the drone closer, as a tree blocked him.

Mina was most definitely hurt, but she was still operational. She crawled around to the sidewalk side of the Suburban, then to its rear, kicking off her coveralls before coming up from behind the men with her service weapon drawn. George had clipped or injured her leg when she dove, because she favored one side.

From above, Leonardo was calculating moves, fearing bullets were not something he could stop.

"You men are impeding a police emergency. Move this truck!" she yelled. Spiro was not a tall man, and was shorter than Mina, but he boldly stepped forward and yelled, "You're no police lady, lady! No way!"

With her weapon drawn, she kept all four men in check and made one final order, "Get this truck out of here!"

Just then, a loud whistle blew from behind the large old tree.

The group could not see the source, but Mina looked in that direction with everyone else. An older African American woman stood on the porch with a whistle in her hand, and it hung from her neck. Mina had made a significant miscalculation.

The safety of the neighborhood watch was due to its captain, Yvonne Jackson. Mina had parked in front of ground zero of the most steadfast woman in southeast Washington, D.C. The whistle was a signal that directed everyone within ten houses in both directions to come outside, and they came prepared.

Yvonne had a booming voice. "Spyro! Is that you?" she yelled, still shielded by the big tree.

Yvonne Jackson had known Spiro for 40 years, and she famously never pronounced his name correctly. Spiro's voice was distinctive, and he knew her voice also.

"Yvonne? Yvonne, we need help!"

Years ago, when George and Spiro started their food cart, they became famous with all the secretaries within five blocks. There were days when these hard-working women barely had enough money for bus fare, but George and Spiro always fed them.

Some days, just before payday, the women would have little, or no money left for lunch. The Greeks were the kindest people they knew. As struggling immigrants, George and Spiro were compassionate to the point that they gave away any profit, but these women worked hard, and the pair of Greeks respected that. They respected the working pool of secretaries more than anyone, and those women would never forget their kindness.

No one ever tallied the score, but everyone knew George and Spiro's ledger greatly favored the women. Yvonne Brown was one of those women, and after she retired, she would still drive out to Souvlaki, the restaurant, to visit and eat with her old friends.

Yvonne marched down her sidewalk toward the Suburban and came into view. Mina still had her weapon raised.

"You can put that gun down, Missy. These boys won't hurt you," Yvonne said.

"They need to move this truck. They are impeding a police emergency."

Yvonne looked Mina up and down with a big smile, then had a big belly laugh. "Lady, I know every cop in D.C., and you're not one of them. So, you just put that gun down and be on your way."

Yvonne was fearless. A circle of 50 residents was closing around the cars: many with bats and broomsticks. Mina nervously scanned the size of the group. She quickly looked to the ground, trying to locate the detonator. Unfortunately, she didn't see it, leaving only one plausible option. She was a mercenary and needed to live and fight another day. Fighting for a paycheck is different than fighting for a flag. She saw one small break in the crowd, so she hobbled one step sideways, still with her gun drawn, then turned and ran reasonably fast down the street.

Yvonne's grandson, Ellis, was standing by her side. He was the next great hope of athletes to come from the neighborhood. Every four or five years, the community produced another future star. He had just signed a letter of intent for a full scholarship at the University of Maryland as a star recruit for their football team. He was a wide receiver and was fast, so he stepped forward in Mina's direction as she ran down the street.

"I could run her down like a wounded dog, Gram," he said.

Yvonne was iron-fisted when she gave orders. "You'll do no such thing!" she put her hand on his big shoulder. "You're safe here with me, and that's where you'll stay."

Leonardo's job here was done, so he lowered the drone, tipped its propellers, then turned and sped back to Gunston Park. He raced across the Potomac River into Virginia, making extraordinary flight moves, which was exhilarating. He sounded like a child riding his first amusement ride with Dart and April watching.

"Who is flying my drone?" Dart asked.

April paused for a half-second. She considered the fortunate outcome and this fantastic invention. She knew, at least for the moment, that things would be okay.

"All is well. He's an old pilot," she said.

The drone sped onto Gunston Park within a minute and dropped gently at Dart's feet as the battery died.

Things weren't over at the Capitol. Jack Hess grabbed a few D.C. police and surrounded the panel van. Marko and Vasko were ready to make a getaway, but it never happened. They were too big to run, and the van was boxed in. Ivan, on the other hand, grabbed a handful of marbles. Hess told all the officers not to fire. "Smart move," Ivan yelled as eight to ten marbles filled both his raised hands. Then, he threw the marbles in his left hand high into the air.

"Everyone back!" Hess yelled.

As everyone retreated, Ivan set off running toward Union Station. He was 50 yards away before any officers could give chase. He turned and faced the officers chasing him just before he entered the train station. He threw two paintballs in their direction, causing them to take cover. Then he turned and disappeared into Union Station, running toward a rear exit near the tracks that would provide him an escape.

So, that left the lonely, odd couple in the park. Dart had his drone back in one piece, but he had a chronology of events that would be curious to anyone asking questions.

Leonardo resumed his counseling in April's ear and said, "I have advised George and the others to vacate that area. It would not be to our advantage if they became embroiled in a local police matter. The woman with the whistle will call the police, and the police will undoubtedly contact the F.B.I. once the white vehicle is linked to the Capitol attack. Your focus should be Mr. Temple, and your best course of action is to contact your Assistant U.S. Attorney for her help."

Leonardo had packed so much intelligence and knowledge of April's world into that statement that she didn't know how to break it down. Where did he possibly gain so much insight into her universe?

April turned to Temple and said, "Okay, here's what we will do. I am going to call my friend, the Assistant U.S. Attorney. But first, I will clarify that you have assisted me by providing critical information to stop an attack on the Capitol. Second, I will ensure they do not charge you with any crimes. Third, and finally, I will make sure you are home, sleeping in your bed tonight. You've done a great service to the country and your friends. Unfortunately, a chemical weapons attack was initiated at the Capitol tonight, which would have endangered the lives of your friends. We thwarted the attack. Your friends are safe."

Dart was uneasy. "Yeah, well, I need an attorney with me."

April supported him, "And you shall have an attorney with you. Give me a moment to make a call."

April called Sophia King, who was already en route to the Justice Department.

"The A.G. called us all in. Traffic is a mess. Every road seems closed, but I'm getting through," Sophia said as she looked at another barrier ahead.

Then April started to knit a story together. "I have a subject with me who was critical to stopping this attack, and—"

Then, Sophia cut her off. "Just come in. Bring him straight in."

That was odd. April hadn't mentioned any gender. There was a long, one-second pause.

Sophia continued. "I have more intelligence than you know, and I know how best to support you. I need you and Darren Temple to get straight to my office. Don't stop for anything or anyone. Just come straight in."

Only one thing could explain the Twilight Zone that April had found herself in, but Leonardo was not in her ear. As much as she wanted to summon him, she needed to focus on Temple.

"Okay, so here's what we are going to do. We'll drive downtown and meet with the Assistant U.S. Attorney. We will call your attorney on the way. They can meet us there. The A.G.'s office will craft something to your liking." April was over-promising with reckless abandon, but she desperately wanted Temple and herself in front of her friend.

Temple reluctantly agreed, packed up his drone, and walked to the car across the street with April like a stubborn old mule being pulled into service.

10

A HERO'S WELCOME

The Oracle of Delphi

April and Temple climbed into her car for the drive to the Justice Department. The fastest route for the six-mile journey started through the adjoining residential neighborhood on Arlington Ridge Road. Temple chuckled as they drove down the street.

"What's so funny?" April asked.

Temple nodded his head to the right. "That's my house."

These were lovely, well-maintained homes, and it was a quiet neighborhood.

"That's quite a contrast," April said softly. Temple understood the irony.

She jumped onto Route 395 toward the District, and headed for the 14th Street Bridge, just past the Pentagon. Traffic was completely stopped. Nobody was getting in or out of Washington, D.C.

"Maybe we should do this tomorrow," Temple said, looking upon the sea of red taillights.

April laughed, then turned on her flashing grill lights and placed her F.B.I. placard on the windshield. She pulled over to the median lane and deliberately worked her way across the bridge into the District, where a barricade stopped them.

Officers wearing tactical gear occupied their hands with automatic weapons and approached the car. April showed her credentials, then asked which roads had been shut down completely. Everything had an eerie, dystopian look to it. As far as the eye could see, the entire District was vacant.

"Looks like Mina made a big mess of things," Temple said, looking down an empty street to his right.

"She failed," April countered. "And you should feel proud that you helped save lives tonight." Her words didn't move Temple's needle.

Temple called an attorney, presumably *his* attorney, who had likely helped him with previous skirmishes. He gave a tortured explanation and asked the attorney to meet him at the Justice Department. A meeting at the Justice Department was no small matter. No doubt, the attorney was equally nervous and excited.

"Let me have someone pick him up. He'll never get into the city tonight. Get his address," April said as she drove across the mall. The Capitol building, now off their right side, was ablaze with emergency lights.

The Justice Department was situated between Constitution and Pennsylvania Avenues, about eight large city blocks from the Capitol. It was also across the street from the vaunted F.B.I. Headquarters.

Emergency vehicles raced through intersections—literally in one-second intervals. Another barricade prevented any traffic from moving forward in front of the Justice Department building. This barricade was different; it was the F.B.I. tactical squad, and they weren't letting April pass without authorization.

The leader of this small tactical team made a call upstairs, seeking permission.

"Affirmative" could be heard on the radio.

April was instructed where to park after bomb-sniffing dogs and agents with mirrors checked out the undercarriage and trunk of her car. They entered the lobby and received credentials but had to wait for an escort. April had been here many times before, but this felt like walking into the belly of the beast for Temple. Even though the "good guys" kept America safe that night, he had issues that this one great heroic act could not mollify.

The elevator door slid back, and the Assistant U.S. Attorney, Sophia King, emerged. April and she were very friendly, but they kept things businesslike in front of Temple. April could barely restrain herself. She had many questions, but Temple was the priority.

Sophia King introduced herself. "The Justice Department and the people of America thank you, Mr. Temple. And thank you for coming down tonight."

Temple wanted to be obstinate and cranky, but it was a beautiful sentiment, and he couldn't help but say, "Thank you."

"We will go upstairs," Sophia continued. "I have arranged a private conference room for you and your attorney. He should be arriving in 20 minutes. Once he arrives, we will outline our next steps, you two will consult, and if all goes as planned, you'll be home in less than two hours with our gratitude. Do you have any questions at this time?"

He shook his head. "No."

The narrative Sophia King had just shared was foreign to April. She had never been involved in a case where the Justice Department made such unqualified promises. Now she had more questions.

They went upstairs, where an aide introduced himself to Temple and escorted him to a glass-walled conference room. Finally, alone, April started in on Sophia King. Sophia motioned to a seating area with a couch and chairs just before April laid into her.

"Let's sit down."

Now sitting, April wanted to question Sophia like a suspect. "Okay, so who are you, and where are we?"

Sophia laughed a little. She knew all this confused her friend, but all the answers would be complicated.

"I'm going to walk you through all this. If you give me a chance, I think my chronology will help you understand."

April gave her a wave to begin, then got up to get a bottle of water from a small refrigerator nearby. She grabbed two bottles and placed one in front of Sophia.

"I'm going to start at the end of the story before I move forward from the beginning," Sophia said. "We don't want Temple talking to anyone about anything, so everything Justice does tonight will support that goal."

April received that odd news like someone had just laid magic hands on her, making a gash in her stomach instantly heal. This unexpected and wonderful proclamation solved a multitude of problems for her. She tried hard not to react, but her eyes gave her away. Fortunately, Sophia was her friend. She couldn't wait to hear what was next.

"I'm still at the end of the story, and I'll work my way back to the beginning in a moment. The State Department is engineering everything we do tonight, not the Justice Department."

That odd news made no sense at all. April's mouth was half-open, ready to fire off a question, but Sophia settled her by gently raising her hands.

"Now, I'm going back to the beginning, and we'll finish with my last statement about Justice."

April was incredulous and bid Sophia to continue, saying, "Girl, you can't throw a ball that far, but give it your best shot."

"Okay, so the beginning of this story starts with Leonardo Da Vinci's call to my father at around 7:45 p.m. tonight."

April played coy. "I'm sorry, who called your father?"

Sophia took a deep breath to slow April down. "It's okay, April. This is all okay. You're going to be fine; just let me explain."

April refused to let her continue. "Wait, wait, wait! Is there some Greek thing involved here?"

Sophia King had grown up in Washington, D.C. went to the University of Pennsylvania for her Philosophy degree, and then stayed there to attend Penn Law School. She met her future husband, Frank King, who was in medical school there, and they returned to D.C., where Sophia went to work for the Justice Department eight years ago. She had grown up as Sophia Popoulos, daughter of Costa and Evanthia Popoulos, and obviously had a familial connection to the Greek Embassy.

"Temple's attorney will be here in a few minutes. We have to wrap this up quickly. There's no way I can go through the whole story. So, listen, and we can follow up tomorrow. I'm going to be here all night," Sophia said. "This whole thing sounded weird when my dad told me about Nick Pappas' technology a few days ago. But after talking with the great Leonardo Da Vinci, I had the same sense of familiarity as my father."

April continued to attempt to play coy. "So, you're telling me you talked with..." she paused. "You're saying, Leonardo Da Vinci? You realize that sounds crazy?"

"It's no crazier than Leonardo Da Vinci flying Mr. Temple's drone a couple of hours ago to help stop a chemical and weapons attack on the U.S. Capitol."

Checkmate Sophia King. April rubbed her eyes. She opted to say nothing.

"The world of International Diplomacy has moved at lightning speed over the past couple of hours, which is why the State Department is running this whole show. The Assistant Secretary of State has requested special consideration from Justice, citing National Security. Justice will ensure Mr. Temple is incentivized to say nothing to anyone."

Sophia paused, then continued, "You, Nick Pappas, and others will meet at the Greek Embassy with their Ambassador and the State Department tomorrow."

April was painfully confused; she had forgotten half her questions. "Why am I going to a meeting at the Greek Embassy?"

Finally, Sophia gave a straight answer. "The government of Greece wants to thank you personally, and the State Department wants you to be there."

April burst out with a muffled scream. "Good God! Can you say one thing to me that makes sense?! Why does the government of Greece want to thank me?"

Sophia had insights, but her duty was to her job, and that job did not involve diplomacy. Instead, it involved getting Darren Temple off the radar. "Look, I'm sorry all of this is complicated. I get it. You've had faith in your friend, Nick, and you've had faith in the technology he developed. Keep the faith tonight. You'll have a clear view of things by tomorrow evening."

April shook her head slightly as if to say no. "I have a pretty big case to drive forward tomorrow. I'm sure I can't make it to that meeting."

Sophia laid everything on the line to wrap this up. She needed to go to Temple's conference room. "The Bureau wants you to be there because Madam Attorney General has already promised State you would attend."

April could not help but smile. "So, you're telling me your boss, the Attorney General of the United States of America, wants me to go to a meeting at the Greek Embassy tomorrow, and my boss knows about it?"

Sophia affirmed the statement, then added, "Yes. But, of course, in addition to your boss, the Director of the F.B.I. knows you're going as well."

April leaned back in her chair and flicked her hand in the air as though implying this was a common request.

"Sure," April said, "make the kids breakfast and get gas in my car. The Director of the F.B.I. has asked me to attend a meeting for the State Department. It's just another normal day."

April stopped to process, then added, "Whenever you open your mouth, you take me into another dimension of confusion. I'm lost! I'm literally lost trying to figure all this out."

Sophia finished with true fireworks. "When I meet with Temple and his attorney, you're meeting with the A.G. to run through things. So, you better get your game face on."

"You're telling me I'm meeting with the Attorney General?"

"Yep, in five minutes."

"I can't!"

"Why?"

"I have to call all my bosses."

"She already did that. She calls one person, and it's all taken care of. We must get to her office. So, get it together, my friend."

The pair got up to leave, despite a million unanswered questions. April could only imagine Gavin Reynolds' shock when he heard she was meeting with the Attorney General and the Assistant Secretary of State.

Temple's attorney arrived. He looked like a cross between a Tommy Bahama advertisement and a Maître d' for an Easter Sunday brunch. He was dressed in all white. His pants, shirt, sports coat, and hat were wrinkled, like they had been worn or hung up in ages. Even his old shoes were once white. He and Temple seemed perfect for one another.

Sophia opened the door to Temple's conference room and turned to April before she walked in. "Just down the hall there. She's waiting for you. Trust me, April. This is all going to be fine."

"From your mouth to God's ears," April said as she looked down the hallway and began her walk into history. The humor she found in Gavin Reynolds' predictable indignation was a pleasant momentary diversion. The Attorney General had a few administrative staff outside her office. They asked April to take a seat.

Her mind raced with many unanswered questions, but she focused on the important meeting. She wondered if the A.G. knew about Leonardo, also. Finally, she settled in on her strategy: answer the questions that were asked, and answer affirmatively if given instructions or orders. No editorializing. No questions.

⸻ ◆ ⸻

Leonardo had shepherded Nick, George, Spiro, and Saruha away from the scene just as he was flying the drone back to Temple. They quickly got back to the boat, shut everything down, then got in their two cars and raced across the Memorial Bridge, back into Virginia, a few minutes before all the streets in D.C. were shut down.

They'd agreed to meet at the Silver Diner in Arlington, but all promised to tell their families they were okay and say nothing of their conquest until they huddled over coffee. Their stories were identical—they had magically left the boat before anything transpired. They got to the diner in ten minutes. Standing by his car, Nick finished his call with Elaine, but the others were still in the truck. Spiro motioned for Nick to climb in.

The three sat motionless as their friend, Costa, explained some instructions.

Spiro spoke up, "Nick is in the car now."

Costa continued, "Oh good, hello, Nicolas. Are you okay?"

"Yeah, we are all good."

Then Nick stopped and whispered to George, "What does he know?"

George laughed as loudly as a crowd of 100 people. "Nicolas is asking what you know, Costa!"

Spiro interjected, speaking fast, "Costa knows everything. He knows about the fake cop lady in the street, how we stopped her, and that Leonardo was flying the drone."

Anxiety raced back into Nick's body. With all the excitement, he had forgotten to reconcile the men's relationship with Leonardo.

Costa resumed talking, "Nicolas, we will meet tomorrow, and all your questions will be answered."

"Meet about what?" Nick replied.

"The heroic act, stopping that attack... What you four did tonight has significant diplomatic implications. It's too complicated to explain now, but you should know that the governments of Greece and the United States are grateful. The Embassy of Greece is hosting a small ceremony to thank you tomorrow."

Spiro was excited and piped up again, "We get to go to the Embassy. All of us!"

Costa continued, "Joining us tomorrow will be the Assistant Secretary of State for the United States, other State Department members, and the F.B.I. Assistant Director. April Falk and her family will join us as well. We also want your family to attend." George and Spiro were so excited; they giggled and hit each other's arms like two small boys.

Nick stacked his concerns, one over another, but his thoughts pivoted immediately to April. He wondered how she would react to Leonardo's outreach to his friends during the last week. Costa had many things to do that night and was anxious to wrap this up.

"So, the ceremony will begin at 1:00 p.m. tomorrow at our Embassy. We look forward to seeing you, and I assure you, Nicolas, you will go home tomorrow evening with answers to every question."

With that, he signed off.

———◆———

The four friends huddled together closely in a booth inside the diner. A waitress brought cups and coffee, and the men all fell silent. They only resumed talking after she left. As was often the case, George was in charge. "We say nothing about tonight until we meet tomorrow at 1:00 p.m."

Nick cared less about the next day's meeting and more about Leonardo's familiarity with his friends.

"So, you guys talk? Do you talk a lot?"

Spiro was quick to reply, "We play backgammon, have coffee, and talk about Olympia."

"Wait," Nick said, realizing Backgammon often included Costa, "Does Costa know him also?"

"He does, and the two of them get along quite well," George replied.

Just then, April called Nick. The men could hear some of what she said.

"Where are you?" April asked.

"I'm at the Silver Diner with George, Spiro, and Saruha."

"Did you all get out of the District okay? Any problems?" she asked.

Nick was confused. How would she have known they were even in the District? But he didn't want to reveal too much for fear he'd be in trouble.

"No, we are all good. Thank you for asking."

April wanted to speak with Nick alone. "When are you heading home?"

Nick slid some money onto the table. "I'm going home now. I'll be there in 20 minutes or so. You?"

"Same. I'll come over," April said curtly.

Nick didn't like the sound of that. It felt like he would get grilled, but once again, he needed more information to provide answers. The men left, walking close to one another, and parted ways in the parking lot.

As fate would have it, Nick and April made the last three turns home following one another. April stopped three houses short, pulled over, and got out of her car. Nick pulled up next to her. She climbed in and told him to pull over. His stomach knotted up, and his heart raced a bit. They both had a bag full of secrets and an equal number of questions.

"What you guys did tonight, stopping that woman, was brave. Are you okay?"

Nick assumed April had received a briefing.

"Well, you react. You don't plan for such things," Nick said, thinking he was being honest, but not saying too much. But did April know they had driven away and not provided a Police report?

Then her next question was off the topic. "So, do you know about tomorrow's meeting at the Greek Embassy? Do you know what it's all about?"

Now Nick knew there was some parity, because he had no idea either.

"No. I got a call from Costa, but it doesn't sound straightforward. Stuff I'm sure I don't understand about International Diplomacy. Are you okay with going? I heard you will be there."

April fidgeted with the air vent. "Well, I have to go, because everyone short of the President told me so tonight, but I don't know why?"

She paused, then continued, "Well, I know exactly why! What am I saying? Have you talked with Leonardo?" Nick shook his head and said nothing, looking down at his steering wheel.

April was half-agitated and half-joking now. "Does he have a phone number yet? I mean, how do I ring this dude up? He seems to drop in on me, and I'm just now realizing that!"

Nick chuckled nervously. "Come to think of it. He's the same with me."

April held her phone up in front of her face. "Hello! *Boun giorno*!"

Nothing happened. Nick thought ahead, planning for the next day.

"Can we drive together?" April was happy to change the topic.

"Sure, but we need to take your minivan. Reggie and the boys are coming with us."

That was new information to Nick.

"That's cool, isn't it?" Nick said cheerfully. April was less enthusiastic. "Well, I have no choice in that matter, either. The Attorney General asked me to bring my family, so they're coming. We should leave at noon."

Nick agreed, relieved that she didn't have more questions for him. She started to get out of the car, then leaned back in.

"Look, all the details of what went down tonight... it might be best not to provide a full account to Elaine. I know it's a big ask, but for the moment, less is more."

Nick would have agreed to anything to get her out of the car. "Absolutely. Say no more." He was still unclear about how much she knew and how she knew it.

⸻ ◆ ⸻

Around 8:15 a.m. the following day, with their parent's approval, Chance and Collin Falk banged on the Pappas' back door to celebrate a day off school with their friends. Reggie soon followed. Elaine made all the boys pancakes and waffles. Reggie sat in like one of the gang, eating his pancakes and the leftovers of his boys', then he had another serving.

Nick came out of his office with a renewed spirit. He had spoken with Leonardo, who had provided him with some answers—the biggest being that April watched the confrontation with Mina via the drone video. Unfortunately, Nick didn't have time to get a full debrief of April's reaction to the great reincarnated man of history playing backgammon with the others.

April had a similar conversation with Da Vinci and Freud from deep within her closet. They assured her they did not know the nature of the meeting at the Embassy but confirmed that they had contacted Costa Popoulos and had solicited his help the night before. Since everyone in her universe praised her, she was less concerned about having concealed so much from her F.B.I. team and leadership. But neither Da Vinci nor Freud knew how the meeting would play out later that day.

Elaine had leaned on Nick quite a bit for an explanation as to why she, Nick, and their boys were going to a meeting at the Embassy. It was less of a concern for Reggie, since April was always involved in government affairs. There was a lot that Nick didn't know, so he was being mostly honest when he told Elaine they'd all have to go to find out. Also, it was less of a concern since Costa was involved. By 11:30 a.m., the boys were all showered and dressed like it was Sunday. Nick cleaned a few things out of the minivan, and they all climbed in and headed for Embassy Row.

—◆—

Every Embassy in Washington, D.C. was majestic. Embassy Row was one of Washington, D.C.s' most remarkable sights. Greece's Embassy was across the street from Ireland's, next to Paraguay's, and, ironically, three blocks from Jo's Café, Mina's old hangout. Greek flags waved slowly in the breeze. In front of the Greek Embassy, Massachusetts Avenue was lined with black government SUVs. Men and women wearing dark sunglasses stood watch over the fleet. Some important people were in the Embassy. The small turnaround driveway, however, had space, and one of the agents watching over the fleet directed Nick to park just behind George's enormous pickup truck.

George, Spiro, and Saruha had been there for two hours. The Embassy staff knew they were coming, but not quite so early. The trio sat patiently and proudly in the lobby. Any little thing that happened seemed like a world event to them. A maid would walk past, and they'd sit up at attention, hoping to be noticed. Two staffers passed by, and they were certain that the world news was soon to follow. Spiro had commented about the beautiful art and finishes.

"This is the history of Greece," he said proudly. George's feelings were stronger.

"This is the history of the world, Spiro," Saruha agreed.

For these three men, immigrants from another country, sitting in the lobby of their home country's Embassy was their proudest day. Over the years, Costa had included them in some events, but they had never been in this lobby. Today was different.

An Embassy staff member escorted Nick, April, and their families into the lobby, where the three proud men welcomed them. The adults talked in hushed tones, but the boys were quite loud, which earned their parents' scorn. Less than a minute passed, and doors adjoining the lobby opened. Costa led a procession of younger staff members, followed by men and women who must have been dignitaries.

A gentleman spoke in English, then Greek, to the crowd, inviting everyone to another room for refreshments.

Costa intercepted the caravan and spoke to the families and his good friends. "May we borrow Nicolas and April for ten minutes? Please—" he pointed to another door, "—if you'll join me in this room."

Nick had never felt so special in his life.

———— ◆ ————

Costa opened the door, and a small delegation of men and women turned to greet them. April immediately recognized the Assistant Director of the F.B.I., Thomas Sherlock. She had met his boss the night before, but he was busy laying the table for what would come with the Justice Department.

Next, another woman in the group introduced herself as the Undersecretary of State for Arms Control and International Security. That was a mouthful, but it was a puzzle piece for April. Then, everyone in the room—including the Assistant Secretary of State—faded into the wallpaper as Costa introduced the Greek Ambassador, Olympia Karatonis, to Nick.

She could not have been more gracious and welcoming as she greeted Nick in Greek, then April in English. "Mr. Pappas and Special Agent Falk, the countries of Greece and the United States of America are grateful to both of you."

Costa excused the entire delegation so that they could join the others, leaving Nick and April alone with the Ambassador. The choreography was stunning. On cue, staff members appeared with a beverage cart at a nearby sitting area.

Ambassador Karatonis graciously invited Nick and April to join her. "Let us sit down for a few minutes before we rejoin the others."

The staff left immediately after coffee and tea had been served, and Ambassador Karatonis began painting a backdrop that answered a myriad of questions.

"I won't bore you with silly diplomatic jargon, but some context will help you understand why you are here with me now. The government of the United States has long wanted greater access to Greece's deep-water ports in Crete because of the proximity to conflict areas in the Middle East. We signed an agreement with the United States last night to grant that access, which is one of the reasons why the dignitaries from the State Department are present. Today is what you might call a quiet celebration of that agreement."

April nodded slightly, as it reconciled her observation about the Undersecretary's presence. However, it didn't explain why the State Department drove the F.B.I.'s moves on this conspiracy.

The Ambassador continued. "Mr. Pappas, may I call you Nicolas?"

Nick could barely say yes fast enough.

"And Special Agent Falk, may I call you April?"

"Of course," April said cheerfully.

"Nick, you have opened the most significant door of our generation with your remarkable technology." Nick froze slightly. April was similar, she was just less obvious.

"Your technology is safe with Greece. We want to provide you resources, security, and the collaboration you need to advance this technology for the benefit of everyone in the world. We are the Centurion Leonardo Da Vinci has spoken of."

That felt and sounded like a lightning bolt. It was one thing to have three old backgammon players talking about Leonardo, but the Ambassador of Greece! The Ambassador had answered many questions, but April and Nick hoped she had more to say. She did.

"The government of Greece has asked for a small favor in consideration of the agreement for the ports in Crete. We have asked that you, Nick, and your friends be removed from the prosecution of the attempted terror attack last night. Your involvement in that prosecution would lead to an investigation that would trace back to your disclosure to April on Easter Sunday. It is for everyone's benefit that your technology remains off the grid while you continue your research."

That solved many of Nick's problems, but April felt vulnerable. Everything the Ambassador just said implicated April in numerous violations of F.B.I. policy. The precision of detail the Ambassador had was stunning. It could only have come from one source.

The Ambassador turned to April. "You have made a series of executive decisions that balanced the need for discretion, protection, and defense of your country. Assistant Director Sherlock will meet with you shortly to commend you for your meritorious service, and he will discuss your promotion within the F.B.I.. But forgive me, it's not my place to disclose all this."

Such was the journey into the bizarre new world with Leonardo Da Vinci. April received this odd news with guarded delight, but it was weird coming from a diplomat from another country. And why was she promoted?

Still addressing April, Ambassador Karatonis had more to say. "I share that information with you because it only makes sense when you understand the genesis of this commendation. This technology will benefit everyone in the world if it is appropriately developed. You are a vital part of a small team, so the government of Greece has requested your engagement over the next 30 days to assist Nicolas. Our Prime Minister personally spoke with your Secretary of State last night."

April could no longer wait to ask a few questions. She composed herself. "May I interrupt you for a moment?"

The Ambassador bid her to proceed.

"You mentioned Leonardo Da Vinci. Have you interacted with him?"

The Ambassador replied, "I have, yes."

"Do the people of Greece have some special connection to him?

"Some Greeks seem to have one, yes. However, Costa seems to have a better understanding of this."

"And you say my government wants me to help Nick for 30 days?"

"Yes."

"What about this case I am investigating?"

"Assistant Director Sherlock will discuss that."

"Does my government have full knowledge of Nick's technology?"

"The United States has focused on this National Security agreement."

April was locked in. "I apologize, respectfully; I don't think that answered my question."

"We made it clear we wanted to develop Nicolas' technology and promised a review sometime within a year. They were quite pleased and focused on the port agreement."

April knew to drop the matter. However, the Ambassador honored her question and follow-up. Did Greece snooker the United States of America? Well, it sure sounded so to April. 'Who would make that deal? What do they think this is? A kids' game application?' she thought to herself. So, April accepted the decision and moved on.

"When do you want me to start helping Nick?"

"Within the next week. And we would like you all to go to Greece on a cultural tour."

Well, that caught Nick's ears. "You want us all to go to Greece?"

The Ambassador affirmed it. Nick looked at April, but she was processing things.

"What are you thinking, April?" the Ambassador asked.

She spoke quietly. "I admit I am confused."

"In the world of international diplomacy, countries often share resources and personnel where there is a mutual interest," Ambassador Karatonis assured April.

April had one request. "I'd like to speak with Sophia King. May I do that before I answer?"

The Ambassador called for an aide and asked him to find Sophia. "She is here today, so she will be right in."

A minute later, an aide escorted Sophia in. She was like a sparkling oasis for April. The Ambassador asked the aide to take April and Sophia to her office so they could talk.

"I have some matters to discuss with Nick in the meantime. We will be right here when you return."

The Ambassador's office was impressive. She had a refined sense of taste, and her office conveyed that.

"Take your government hat off and talk to me like you're my attorney for a minute," April said as she jumped into her questions. "Am I in some kind of trouble?"

"People in trouble aren't asked to help on a diplomatic mission, April," Sophia settled her. "I told you last night, all will be okay."

April was dubious.

"If I start listing the number of laws and Bureau guidelines being skirted, we'd be here until tomorrow."

Sophia was quiet and reassuring. "Imagine this as a card game—that's the best analogy. National Security trumps all others. Our government wants access to those ports in Crete, and it's a small matter to have one agent help a friend on a cultural tour for 30 days."

April wanted another bite at the apple. "It may be a small matter to help, but does our government know what they traded?"

Sophia spoke plainly. "Our government places a priority on those ports."

April laughed, but not too loudly. "That's what the Ambassador just said! Is everyone else going to give me the same answer?"

Sophia said nothing, but the look on her face implied it was time to move on. April closed her eyes and focused.

"But why am I getting some kind of promotion, and why does the government of Greece know about it?"

Sophia looked around the office and motioned to all the Greek art. "I would have thought the Ambassador told you that the Greek Prime Minister called the Secretary of State last night."

April was quick to reply. "She did."

Then Sophia wrapped it all up like a tiny box with a beautiful ribbon. "When a world leader calls the Secretary of State of the United States of America and says, 'you can thank April Falk for opening this important door,' don't you think your stock shoots straight up? The Secretary of State called the Attorney General and the Attorney General called the Director of the F.B.I.. If this were poker, you'd be holding a royal flush. You have paid the highest dividend possible in service to our government. Face it, girl, you're a star! That's why all this is happening. Our job at Justice is to make all this right, prosecute these guys, and find out who is behind all this. Your job is to spend 30 days in Greece and be nice to our outstanding friends."

It was weird, but that made sense to April. She raised her shoulders, pursed her lips, and said, "Okay. We shouldn't keep all these people waiting."

They opened the door, and the aide escorted them back to Nick and the Ambassador. Nick had a look on his face that April had never seen.

"What's up?" She whispered as they all left to rejoin the large group.

Nick could barely talk. "Uhm, I just had a strange conversation with the Ambassador."

April laughed, "It couldn't be weirder than the one I just had!"

"Oh yes, it could," Nick said. "Just wait."

Then they were absorbed by the crowd.

The large room buzzed with conversations. George, Spiro, and Saruha had not spoken English in 15 minutes, as they talked to every staff member possible. Then, finally, Ambassador Karatonis approached a podium in the front, and everyone fell silent. Elaine leaned into Nick.

"What's going on here? Those people from the State Department implied you're a big deal. Why would they say that?"

Nick wasn't sure how to respond, but said, "That project I was working on with the guys in India has been well received, and the Greek government wants to help."

Elaine was amazed, proud, and confused.

"Let's listen," Nick said, motioning to the Ambassador.

"May I ask Mr. Nicolas Pappas to join me?" Nick looked at Elaine.

She was so proud. She gave him a loving push on his back. Then, everyone parted as Nick walked forward and stood by the Ambassador.

"The governments of the United States and Greece wish to thank you for your work in Science and Technology. The government of Greece confers upon you, Mr. Nicolas Pappas, the title of Honorary Consul to the Embassy of Greece in the Ministry of Science and Technology. Your dual citizenship allows us to make this confirmation. This Consul status will provide you with resources, collaboration, and security as you continue your groundbreaking work. Please, everyone, join me in congratulating Mr. Nicolas Pappas."

The crowd roared as the Ambassador placed an ornate Greek sash over Nick's head, then an Embassy lapel pin on his jacket. Everyone applauded, but George, Spiro, and Saruha cheered from the back like the Twins won another Series. George teared up with pride as he clapped and wiped his eyes.

Nick looked back to Elaine, who was obviously speechless and could barely raise her right hand to her heart to praise him and send him love.

Then he looked toward April. She was standing to Elaine's right with Reggie, whom she had just told about the trip to Greece. They nodded a salute to one another.

April and Nick had spent 12 days in total suspense. Twelve days where every six hours elevated concerns so greatly that they forgot about the calamity behind them. It would only be weeks later that they would realize the enormity of harnessing this artificial intelligence monster. They had just run humanity's first gauntlet and had become the first to prove that you don't control artificial intelligence; you try to keep up.

But, oh God, what if a technology like that was built with an architectural plan different from Leonardo's and Freud's? What if someone manipulated it? What if it was designed by someone with motivations less pure than Nick's?

At that moment, weeks later, they would fully understand Leonardo's most extraordinary teaching: they needed time to develop this safely. They needed a safe harbor. Greece's government would be that port in the storm, with Nick the dreamer, and April the leader, though more time would pass before she fully understood her role. Over 1,000 pathways led to that moment in the Embassy, which Leonardo had targeted and planned for since he and George met.

Nick tried to rejoin Elaine and their boys, but the Greek Embassy staff members who wanted to congratulate and welcome him stopped him. Then, the room fell quiet again as the official business between the governments was announced.

Out of respect, Elaine said nothing until the diplomats finished talking, but wasted less than a half-second before saying, "Okay, what just happened here?"

April, Reggie, and the boys crowded in to hear more. George, Spiro, and Saruha stood behind Nick like his coaching staff.

Nick sounded apologetic and uncertain but gave all of them the great news. "They want us to go to Greece!" He could have blown Elaine over with a puff of air. But, instead, it was a glorious finish line for her as well. All those years Nick toiled in obscurity had finally resulted in someone saying, "Good job," and then awarding him the commission of a lifetime.

April spoke up. "And we're also going." Her boys had heard nothing when she told Reggie.

"Does that mean we are also going?" Chance asked. April smiled, looking at her pride and joy. "We wouldn't go without you."

Costa rejoined the group in its moment of private celebration. "We are organizing visits to several archeological sites and cities, including Olympia and Delphi."

George interrupted. "If you're going to Olympia, we will await you when you arrive."

George wasn't asking. He was telling.

Reggie jumped in. "If you're coming also, does that mean you'll take us to your favorite restaurants?"

"Yes, and they will treat you like a king," George said proudly.

Costa was happy for the group but was extremely busy and needed to reclaim control of the discussion, so he interrupted. "When in Delphi, we will arrange a private visit to the Temple of Apollo."

The Temple of Apollo was better known as the location of the Oracle of Delphi, whose name was Pythia. Nick was curious, because a private visit to one of Greece's most important shrines was no small matter.

"What are we to do there?" he asked.

Costa's answer was vague. "Just go. See what you see. We are curious to know if the Temple reveals anything to you and your other friends."

Of course, Costa was making a reference that included Da Vinci and Freud. Nick wondered what mysteries could possibly be uncovered while there.

George spoke to April and her family, "Delphi is so beautiful. High on a mountain, overlooking the pretty seaside town of Itea and the Gulf of Corinth. My old friend owns the Nidimos Hotel there. We will sit on their balcony and see the best of God's sunsets each day."

Elaine poked Nick with a curious finger.

"Costa just said, 'Your other friends.' Who is he talking about?"

Just then, George, Spiro, and Saruha's phones lit up with a text message. It looked like a spam text, but it was a photo of Da Vinci and Freud standing before the Temple of Apollo. Leonardo looked like an American tourist, wearing non-wrinkle pants and a long-sleeved L.L. Bean shirt. He had a DSLR camera around his neck and was wearing a San Francisco Giants baseball hat.

A meme above and below said, "The Oracle's Secrets Await You."

Reggie glanced at the image and pointed at Freud. "Hey, that's the odd online professor teaching me the Electric Slide! He's hilarious. He's got these videos that keep popping up on the computer."

The boys all looked to see. Everyone halted when Chance exclaimed, "Whoa, that's the *Dancing Cowboy*! I know it is. Look, guys!" Chance said to the other boys, who were jockeying for position. "That dude thrashes everyone in the gaming world."

Elaine looked in and focused on Leonardo as well. To her surprise, the man in the Giants' hat was not some gaming cowboy. He had been teaching her to draw on her iPad with an Apple Pencil using Procreate.

"He's not a cowboy dancer, boys."

The boys corrected her, "Dancing Cowboy!" Oh, how they were embarrassed.

Elaine continued, "He produces these great videos that teach people how to draw on an iPad. He's talented."

April muttered, "That guy is teaching you to draw? Well, that's the understatement of the day! Is the Mona Lisa one of the lessons?"

George, Spiro, and Saruha looked like three Cheshire cats, but they said nothing. Then, finally, Nick looked to Saruha with a request.

"Can you make a table for 11 at The Plaka tonight?"

Saruha promised he would without hesitation. Nick looked at April, and they both looked at the group; then April took the lead.

"Let's have dinner, talk about our trip, and then we'll talk about our new friends."

The End

Acknowledgements

I had never read an acknowledgment page before I had to write this one. Welcome if this is your first. The few I read could have been more interesting, so I'll make this one worth a few minutes of your time.

A first-time, unrepresented writer wanders into this role as an author, much like an explorer without prior experience, a map, or a plan. Couple that with a lack of confidence in one's writing ability, and you join me on my first day. I had a story. I had a good story, but advancing it meant jumping off a cliff.

There were many unrelated starting points in this journey. They were random and unrelated. These are only a few that prompted me to begin. The first, many years ago, was the forethought that the potential of A.I. is limitless. That led to some architectural designs and research into data requirements to drive an engine like the one outlined in Pali. Separately, one day, I searched for something to listen to on the radio. The owner and operator of *WABC* radio, John Catsimatidis, co-hosted a live show with local icon Curtis Sliwa. They spoke of the history of grocery stores in New York. It was a vivid highlight reel. The ethos of Pali is history and learning. Then, when touring Greece, I was struck by how the people cherish their rich history. Finally, I wrote this whole story as a screenplay before writing this manuscript.

A thousand strangers listened politely as I shared outlines of this story. The story changed slightly with each reveal, but my ability to hold their attention was constant. If you were one of those strangers, I thank you for enduring me.

A hundred friends listened politely also. They could not have been nicer, but that's what friends are for. Ellie Campbell listened as I previewed the architecture of Pali for the first time. She encouraged me to keep going. So, I cornered three childhood friends in my car as I drove them to the airport one day. Nervously I asked if they'd listen to 10 minutes of this story I was building. They were so polite and encouraging. Thank you, Debbie, Sue, and Alyson. Cathy Winter and her daughter Laura, our good friend, listened to an audio version of an early draft. They made suggestions, and I incorporated them into the story. How can you thank someone enough for that? Finally, my neighbors and friends, John Ward and John Carvelli, generously acted as sounding boards when I needed insights and encouragement.

Anthony Vitolo read the entire first draft of the screenplay I wrote. I don't know if writing a script first is recommended, but Anthony was supportive and polite. Jonathan Koch graciously gave me hours of his valuable time. He asked plot and character questions that exposed weaknesses. I hope he is pleased. I've called upon my long-time friend, Eddie Lauth, with business and creative ideas a hundred times. His answer is always the same, "Let's go. I'm in." When I approached him, he made some calls for me. Another lifelong friend, Mark Baughman, has worked as a federal agent and is now an active shooter expert for television. He has helped me advance the notion of using sound as a deterrent before police arrive on the scene. Finally, my old friend and roommate, Andy Bingaman, was the strongest, most disciplined person I've ever known. He lost his battle with Lymphoma at a young age. He was an FBI agent and started his career in Wichita, then finished in Washington, D.C. In between, he transferred to New York and became friends with (now retired) FBI Agent Jack Hess. So many years later, Jack still answers my questions. My friend, Dr. Mitchell Brin, helped explain memory function to me. It's a complicated topic, but he made it easy to understand.

Many years ago, the Nichols family introduced me to their Greek community in my hometown. The Greek language seemed oddly natural, so I kept up with it occasionally. The Greeks are great people; everyone should visit the Greek mainland as we did a few years ago. Olympia and Delphi may inspire you also.

I wrote 90% of this book in the Newport Beach Public Library. I looked upon the racks and racks of books as inspiration. A library seems different when you're trying to write a book. I was also inspired by countless people of all ages who still go to the library to study, research, and read quietly. Thank you to the library staff and city officials for making this extraordinary building available.

My editors were Sue Moore, Cathy Winter, Laura Sherlock, Colleen Taricani, Joey Taricani, Kate Taricani, Michelle Morgan, Haley Larkin and Katherine Graham. We can all thank them for their patience. Eve Hard was the artist who worked with me to develop the cover. Our

Publisher was Randi Harvey, RH Publishing. Randi generously fielded many of my questions. I drafted an early version of a movie treatment with Brandt Sohn. I hope he sees this. I'm sure this will entertain him.

My parents, brother, and sister have had to listen in for years as I came home with new directions, ideas, and ventures. That must have been exhausting. Thank you for supporting me.

I'm uncertain if I'm the second, third, or fourth-best writer in our family, but I'm confident I'm at least a distant second from my wife, Colleen. She's a brilliant writer and communicator. She's also patient, allowing me the latitude to research new ideas that must certainly seem scattered but occasionally ends up as something wild and unimaginable, like an emerging industry, another new patent, or a published novel. I tell people my world is good if our children are happy, safe, and healthy. They are. Our son, Joey, inherited the best characteristics of his mother (reading and intelligence) and few, if any, of his dad's bad traits. Our daughter, Kate, has a great imagination and great intellect. That's a powerful force. Fortunately for us, she has been involved in the performing arts for years. Watching her and her friends grow their talents on stage has always inspired me. We are grateful to her teachers, fellow actors, actresses, singers, and orchestra members. You have all been superb and have bright futures.

Lastly, thank you to you. I'll always be astonished that I produced something people chose to read. A special thanks to you if you took the time to share this book on your social media and with your friends. Unrepresented, first-time authors who self-publish face daunting odds of being marginally successful. There are no publicists or campaign managers to help circulate the story. And when you're new to an industry, you don't have connections, so it's one book at a time, and the author does their best.

Now, with some limited writing experience, I'll make a plan, and then I'll try to do this again because I know the hidden secret of where Nick, April, and Leonardo are heading.

The journey continues.

About the Author

Pali is Joseph Taricani's first novel. He has had leadership roles in some of the world's largest, most outstanding companies. He has also been a policy expert for Congress and an entrepreneur. He holds many U.S. patents and copyrights. He previously led a team to develop one of the first artificial intelligence systems used in the financial services industry, which was revolutionary. He's been fortunate to work with some of the world's best business executives and government leaders. His happiest days involve exercise, spending time with family and close friends, working with people he likes, and traveling to foreign countries with his family. He also dabbles in languages which include Greek, Italian, and Spanish. Originally from Pennsylvania, he has lived and worked in Minnesota, North Carolina, Texas, northern California,Wisconsin and Washington, D.C., twice, as well as work assignments in foreign countries. Mr. Taricani and his family reside in southern California.